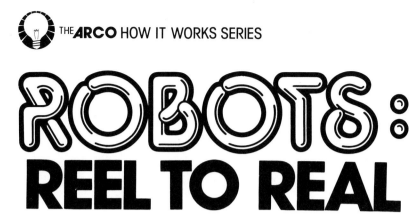

THE**ARCO** HOW IT WORKS SERIES

ROBOTS:
REEL TO REAL

Barbara Krasnoff

ARCO PUBLISHING, INC.
NEW YORK

Published by Arco Publishing, Inc.
219 Park Avenue South, New York, N.Y. 10003

Library of Congress Cataloging in Publication Data
Krasnoff, Barbara.
    Robots, reel to real.

    (The Arco how it works series)
    Includes bibliographical references.
      1. Automata.   2. Robots, Industrial.
I. Title.   II. Series: Arco how it works series.
TJ211.K7   629.8'92   81-3514
ISBN 0-668-05139-6 (Cloth Edition)   AACR2.
ISBN 0-666-05141-8 (Paper Edition)

Printed in the United States of America
10 9 8 7 6 5 4 3 2 1

11/83

To Robin, who got me the job;
To Bob, who got me the book;
And to my parents, who got me.

Contents

Chapter One

Unreal Robots

IMAGINE

PICTURE THIS: MARY and John Doe are sitting in their futuristic, 21st century plastic living room, watching a space opera on their Tri-Vid. John decides that he'd like a can of artificial beer, and presses a small button on the coffee table. Almost instantly, a silver man with glowing eyes steps softly into the room, bows debonairly and says in a pleasantly modulated tone, "How may I serve you, master?"

Now try this: A cute, cone-shaped metal machine rolls through the crowd at a busy shopping center. Multi-colored lights flash around its gleaming circumference and a wide smile is painted over the speaker on its "face." Occasionally it will stop in front of a bemused pedestrian, shake hands and announce, in a high-pitched, electronic voice, "Hi! I'm Herbie the robot!"

Just one more: In a long factory room, amid the whine of machinery, squat metal shapes with long, hinged arm mechanisms work busily at their programmed tasks: screwing parts together, welding car doors, stacking boxes, assembling delicate radio equipment. They can't speak or hear, won't even recognize the fact that you're there—you can pound on them, yell at them, curse at them—nothing will divert their computerized attention. Unless, of course, you hit the "off" switch.

The three scenes just enacted describe machines that are

commonly known as robots. However, experts would only call one of them a *real* robot. The other two are imitations.

The mechanized servant owned by the Doe family, for example, is part of a typical 1950s vision of the future as popularized by science fiction novels and films. However, most of today's scientists and technicians agree that, while the abilities of intelligent automatons will make them practically indispensable in the future, serving beer will probably not be one of their functions. For one thing, the amount of programming and sophisticated circuitry that would have to go into a robot with the ability to perform even a reasonable number of household tasks, such as washing dishes or cooking eggs, would be awesome (although perhaps not quite as awesome as its price).

The second scenario featured a machine that, by most definitions, isn't really a robot either. You may have seen one of these roving mechanical toys at a shopping center opening or new car show, beeping and telling bad jokes, and generally *looking* like a robot. But odds are that somewhere behind the scenes there was a technician turning a dial to keep it moving or speaking into a hidden microphone to become the robot's "voice." Not exactly an independently functioning machine. Not exactly a robot.

The third example described real, working robots—the kind that are presently in wide use around the world putting together automobiles, assembling delicate electronic parts, handling nuclear material and exploring the surface of Mars. And while most of these machines look like anything from a high-class vacuum cleaner to something a five-year-old built using an erector set, they are capable of independent, programmed work—they are robots.

Of course, this brings up a very logical question: What exactly *is* a robot? According to some manufacturers, it is almost anything mechanical that does a job. According to some science fiction writers, it is a computerized form of life. And according to some grade-B movie makers, it is an animated tin can that gets its kicks out of carrying off beautiful starlets.

The concept of a robot is not a very new one. In fact, for almost as long as there has been a history, people have fantasized about artificial beings that would be at their beck and call and, unlike human slaves, would not be subject to human laws.

THE FIRST DREAM

Perhaps the first mention of these ideal servants was in the story of the Greek god Hephaestus (or Vulcan), who is said to have fashioned two living female statues out of pure gold. Following his example, the mythical artist Pygmalion sculpted the statue of a beautiful woman, subsequently fell in love with it and convinced the gods to give it life.

A different type of wish fulfillment was represented by the *golem,* a clay figure said to have been shaped by a medieval rabbi to protect his people from their oppressors. The *golem* was animated by certain holy words which, when placed in its mouth (or written on its forehead, depending on the version of the story) brought it to life. It was deactivated by removing the words.

As Europe entered what is known as the Industrial Age, people became more fascinated than ever before with the idea of artificial life. After all, now that they had mastered mechanics, what could humanity not create? Animated mannikins, carefully crafted to mimic human mannerisms and accomplishments, were popular in the richer households. Foremost among the manufacturers of these toys were Pierre and Henri-Louis Jacquet-Droz, who in the 18th century were famous for their elaborately lifelike mechanisms, including a young woman who played the piano and a boy who could draw and write.

In 1769, Baron Wolfgang von Kempelen produced a chess-playing Turkish gentleman. The life-sized automaton sat at a large cabinet with a long pipe in one hand, consistently beating its living opponents at chess with the other. Such notables as Edgar Allen Poe sought for years to uncover the trick behind the chess player. They finally determined that the figure was operated by a small man sitting inside the table, hidden by a trick door.

This interest in artificial life began to appear in literature as well. Mary Wollstonecraft Shelley's classic novel *Frankenstein,* written in 1818, while not about a robot as we understand the term (Dr. Frankenstein's monster being assembled from used human, rather than metal, parts) did reveal a growing public

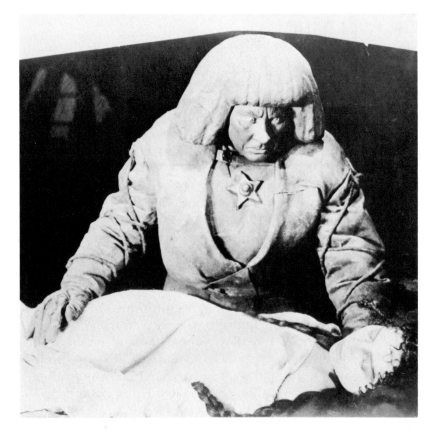

A scene from a silent film version of *The Golem*. According to legend, an artificial man was created out of clay and brought to life by mystical means. (© *courtesy Janus Films, Inc.*)

awareness of the roles that science and technology were beginning to play in the world. It also reflected popular concerns about that technology, which can still be found today. For example, *Frankenstein*'s apparent theme of the unwary scientist invading God's territory by creating life eventually inspired a spate of usually inferior horror films that ended with the doleful phrase ". . . there are some things Man was not *meant* to know. . . ." Author Isaac Asimov later coined the term "Frankenstein complex" to describe this fear of new technologies, especially of robots and their ilk.

While all the aforementioned automatons indicated an almost

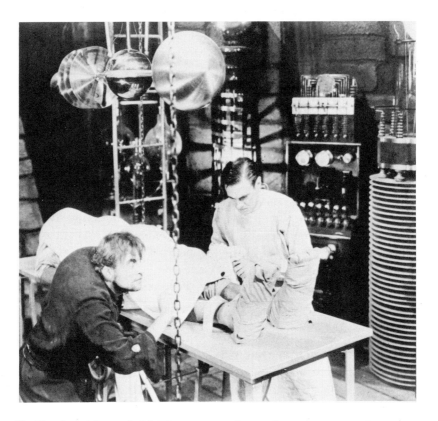

Dr. Frankenstein made his Monster out of spare human parts. Today, scientists prefer using metal. (*From the motion picture* Frankenstein, *Courtesy of Universal Pictures*)

universal interest in the idea of a robot, it took a Czechosolvakian playwright to actually invent the word that would thereafter define our concept of artificial life. In the 1973 edition of the Oxford English Dictionary, for example, the entry for "robot" reads as follows:

> Robot (ro · bot) 1923 (– Czech, f. *robota* compulsory service) One of the mechanical men and women in the play *R.U.R.* (Rossum's Universal Robots) by Karel Çapek; hence, a living being that acts automatically (without volition). b. A machine devised to function in place of a living agent; one which acts automatically or with a minimum of external impulse.[i]

Çapek's popular play concerns an industry in which human-like servants are artificially created out of biological material in order to serve the human race in the factories and the military forces. He called these manufactured workers "robots," from the Czech word *robota*, meaning obligatory work or servitude.

Çapek's robots have the appearance, and all the capabilities, of their masters—with two important exceptions: the ability to reproduce, and what the author enigmatically terms a "soul" (the mathematical formula for which would have been very interesting). The robots—like their cinematic cousins of later years—proceed to take over the world. One lonely scientist manages to create two last robots, a male and a female, both with the missing qualities intact. With humanity practically extinct, and the robots due to wear out in a few years, these two wander off hand-in-hand into the sunset to begin a new and (the author hoped) better civilization.

Obviously, Çapek's robots are only distant kin to their mechanical descendants—they were, after all, created as a social and political commentary rather than as an accurate forecast of technology. But the word passed into common usage, and as it became more and more likely that humanity's artificial servants were going to contain electrical circuitry rather than hearts and lungs, the term came to be understood as referring to a machine.

(Çapek did, interestingly enough, foretell a problem with robots that scientists are wrestling with today—programming.) "They've astonishing memories, you know," explains one of Çapek's scientists in *R.U.R.* "If you were to read a twenty-volume encyclopedia to them, they'd repeat it all to you with absolute accuracy. But they never think of anything new."[2]

SCIENCE FICTION

The creatures of *R.U.R.* were just a first in what soon became a literary obsession. Walking, talking and thinking mechanical robots as we know them became widely popular in the flood of science fiction pulp magazines that began to pour forth in the 1930s and continued into the '40s and '50s. The stories in these publications explored the marvels that a technology-happy humanity was looking forward to in the years to come. Aliens and evil empires galore appeared beyond the stars, mad scientists created havoc on Earth, and stalwart heroes zoomed off in their trusty rocketships to save the universe from assorted dangers. Very often, these heroes were either battling against, or accompanied by, a robot.

The automatons that appeared in the fiction of this time could either be friends or enemies of their human creators. For example, the robots featured in Lester Del Rey's "Helen O'Loy" (1938) and Eando Binder's "I, Robot" (1939—not to be confused with the later Asimov collection of the same title) are essentially happy to be faithful servants of humanity. This is in sharp contrast to the chilling tale "Fondly Fahrenheit" by Alfred Bester (1954), in which a sophisticated robotic servant develops a penchant for murder, making its owner an unwilling accomplice to the crime, since he is considered responsible for his property's actions.

For the most part, these robots were all formed in their creators' image—with arms, legs and even recognizable faces. While this may have appealed to readers' imaginations (indeed, many compliant female robots were created by authors to serve men dissatisfied with the human variety), a human form is actually not necessary, or even desirable, for a robot. Most robots in use today in industry and space are totally functional in design and in no way resemble the people who employ them. (Of course, there are many machines on display these days that *look* like functional robots, but we'll deal with them later.)

This is not to say that all authors wrote about humanoid robots. In 1951, Arthur C. Clarke published a short story entitled "The Sentinel," that dealt with an alien artifact found on the moon. This concept was later expanded to become the movie and novelization *2001: A Space Odyssey* (1968). In the expanded story, a manned spaceship sent out to investigate a signal being transmitted by the mysterious monolith is almost totally supervised and operated by a computer, the HAL 2000. HAL, in fact, *is* the ship, in the sense that the computer itself is the brain and the

In *2001: A Space Odyssey,* the space ship itself was the robot. Here, an astronaut disengages its memory circuits. (*From the MGM release* 2001: A Space Odyssey, © *1968 Metro-Goldwyn-Mayer, Inc.*)

ship the body. Eventually, HAL, following a time-honored tradition (in fiction at any rate), decides that it can handle things better than its human passengers, and begins to "deactivate" them. The one surviving astronaut is forced to perform what can best be described as the robotic equivalent of a lobotomy in order to regain control of the ship.

Another author who suited his robots to their tasks was Isaac Asimov, who is well-known for his short stories featuring robots, many of which were collected into the two volumes *I, Robot* (1950) and *The Rest of the Robots* (1964). In the course of these stories, Asimov describes the rise of a powerful robot industry, the increasing sophistication of its products and the efforts of a robot "psychiatrist," Dr. Susan Calvin, not only to cure her charges of complicated problems with their thinking processes but also to cure humanity of its disposition to regard their mechanical helpers with distrust.

Asimov's robots, while still technically way out of the reach of today's science, are still simply mechanical servants — programmable, predictable and capable of doing only what they were created to do. For example, in "Little Lost Robot," a mining robot, told by an impatient overseer to get lost, does just that. It does not have the capability to differentiate between the phrase's literal and figurative meanings. Of course, as the tales reach farther into the future, the robots continue to progress; eventually, Asimov presents us with a robot so complex that it is indistinguishable from a human.

One facet of Asimov's robot stories have, so to speak, transcended their medium, and can be directly or indirectly found in many other tales dealing with intelligent automatons. These are his Three Laws of Robotics — rules of behavior that are indelibly ingrained into his robots' positronic brains. They act as a set of Golden Rules through which the machines can be controlled and a suspicious public can be assured of its own safety. (If Čapek's robots had been programmed with these laws, for example, their insurrection could not have taken place.) The Three Laws are as follows:

1. A robot may not injure a human being, or, through inaction, allow a human being to come to harm.

2. A robot must obey the orders given it by human beings except where such orders would conflict with the First Law.

3. A robot must protect its own existence as long as such protection does not conflict with the First or Second Law.

While many science fiction readers have taken these laws to be the last word in robotic safety features, Dr. Asimov himself has admitted that they are somewhat ambiguous. What, for example, is meant by "harm"? What is the definition of a human being? Can the robot always know when orders under the Second Law are contradicted by the First? Such ambiguities provide the conflicts in most of Asimov's stories. For example, in "Liar," a robot that can read minds is placed in the awkward position of either lying to humans or harming them emotionally. In the novel *The Caves of Steel* (1953), a humanoid robot detective, in order to stop a riot, apparently threatens the crowd with a gun, thus violating the First Law.

In the science fiction of the 1960s and '70s, many of the robotic stories dealt less with robots as a separate species and more with the complex relationships that can develop between humans and machines. There was much exploring and questioning of the differences between mechanical and biological life. Some of these problems are shown humorously, as in Robert Silverberg's 1971 story "Good News From the Vatican," in which a robot is elected Pope. In a more serious vein, there is Asimov's "The Bicentennial Man" (1976), in which a robot, whose ultimate ambition is to be accepted as a human, finally decides that what separates the two species is the reality of death and sets about ensuring his own demise.

However, for the most part, robots today seem to be playing a much less prominent part in literary science fiction. Perhaps this is due to the fact that, since robots are now becoming a reality, writers are forced to look further afield for believable fantasies. This problem is illustrated by the total impossibility of writing a serious story today about an advanced Martian civilization—we *know* they ain't up there.

FANTASTIC FILMS

But while robots may have waned somewhat in literary popularity, the motion picture industry has taken up their cause with great enthusiasm.

When movies were themselves a new technology, German director Fritz Lang created a powerful political vision of the future that is now considered a classic of the silent film era: *Metropolis* (1927). In the year 2000, according to Lang, humanity has split into two distinct classes: the privileged owners and the

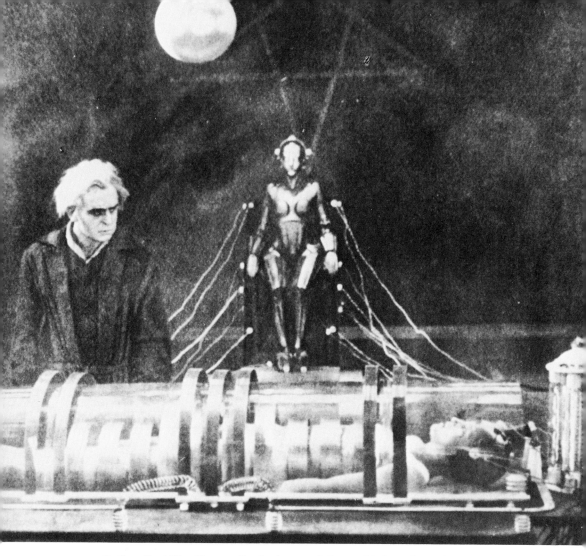

In the silent film *Metropolis,* the image of the sophisticated metal automaton was first brought to gleaming life. (*From* Metropolis, © *courtesy Janus Films, Inc.*)

downtrodden workers, the latter led by the heroine Maria. Upon the instigation of master industrialist Frederson, a scientist creates a female robot to lead the rebellious workers into a premature and therefore disastrous revolt. While that robot is quickly turned, through a rather magical process, into an exact duplicate of Maria, its first appearance is breathtaking: a shining metallic figure that slowly finds itself capable of movement, and through movement, life.

Unfortunately, few of the motion pictures in the decades that followed *Metropolis* equalled it in quality. From the 1930s on, dozens of grade-B science fiction/horror films were released that featured robot monsters as their villains. Such films had titles like *Robot Monster, Gog the Killer, Target Earth* and *The Colossus of New York*, and usually followed one of two basic plots.

The first concerns a robot that was built by a mad and/or absent-minded scientist, who creates the machine for the good of humanity. However, ". . . there are some things . . ." and as soon as the soulless machine is turned loose, it does its best to destroy everything in its path, before being stopped by the fearless hero and the scientist's screaming daughter.

The second plot usually puts the robot in the employ of a tribe of scheming aliens from Mars (or Venus, or you fill in the blank) who, for some strange reason of their own, want to take over the Earth. The machine then goes around smashing through buildings, or sending lightning bolts streaming from its fingertips, or hypnotizing hapless Earthlings into total submission, until— you guessed it—the hero finds some weakness in the robot's makeup that its creators didn't foresee and defeats the alien menace.

Whatever the plot, the end was always the same—the plug was pulled, the heroine was safely in the arms of the hero and everyone was happy—except the robot. However, there were exceptions to this rule.

The Day the Earth Stood Still (1957) is a well-made morality tale concerning Earth's propensity for conflict. The alien Klaatu lands his starship in the United States. He is accompanied by Gort, a 12-foot-high silvery automaton. Their mission is to warn humanity to reach a peace, or else. The other peoples of the galaxy have apparently never heard of the Frankenstein complex; they have turned over all authority to robots such as Gort, who essentially make up the galactic police force. The system is presented as quite logical on the part of the aliens; after all, what could be more just and incorruptible than a robot? (This rather troubling idea is also touched upon in *The Caves of Steel*, in which the robot detective expresses to his human partner the opinion that justice is simply appropriate retribution meted out to those who have transgressed against the established law. He

In a demonstration of Asimov's First Law of Robotics, Robbie the Robot is unable to shoot humans, even on command. (*From the MGM release* Forbidden Planet, © *1956 Loew's Inc.*)

has not been programmed with the subtleties of right and wrong.)

Robbie the Robot, truly one of the most popular automatons in the annals of film and the first in a long line of lovable machines, appears in the 1956 movie *Forbidden Planet.* Unlike Gort, Robbie is of human manufacture, albeit with the help of knowledge obtained from alien sources. It is the handy household robot we've all dreamed about: it can cook, clean, sew and manufacture anything once given a sample (including substances of an alcoholic nature). It also has a voice that, while amusing, is thankfully not given to the sort of wisecracks that afflict modern robotic actors.

Forbidden Planet is one of those films that indirectly refers to Asimov's Three Laws of Robotics. At one point, the scientist Morbius, who built the robot, gives it a blaster and orders it to fire upon a group of horrified astronauts. Faced with a conflict of the First and Second Laws, poor Robbie almost blows a gasket before Morbius cancels the order.

However, for the most part, film and television directors seem to feel that robots are an innate threat and have presented them as such. Many times, the robots are shown to have taken over societies that they once served. In *THX 1138* (1969), humans live in a totally sterile, ordered community governed by a computer, with enforcement by grim-looking law robots with silver faces. In the long-running British television series *Dr. Who* (1963-   ), the timetravelling hero frequently meets up with his most dreaded enemies: Daleks, dome-shaped robots that were originally created as weapons during a planetary conflict. Both sides of that war have long since perished, but the evil Daleks live on, following their programmed orders to eliminate all life wherever they find it.

American television hasn't been very kind to its robotic characters either. Even the program consistently cited as television's one example of good science fiction, *Star Trek* (1966-69), portrayed most of its robots as dangers rather than intelligent tools.

The most common example of this tendency was the story line that concerned a computer and/or robot gone insane. The plot would go something like this: A computer and/or robot that is in some way superior to the usual model is introduced onto the starship *Enterprise*. It looks over the territory, decides that it wants to take over and exterminate all biological life, and proceeds to do just that, over the objections of the crew. The episode ends with the hero, Captain Kirk, using his "superior" human logic to devastate the computer's programming; the machine then self-destructs in some suitably dramatic fashion.

As could be expected, *Star Trek —The Motion Picture* (1980) used this same scenario. A huge mass of energy, that turns out to contain the ultimate robot, is on its way toward Earth, devouring all life in its path. The *Enterprise* is sent out to investigate. Curious about the starship's intentions, the computer turns one

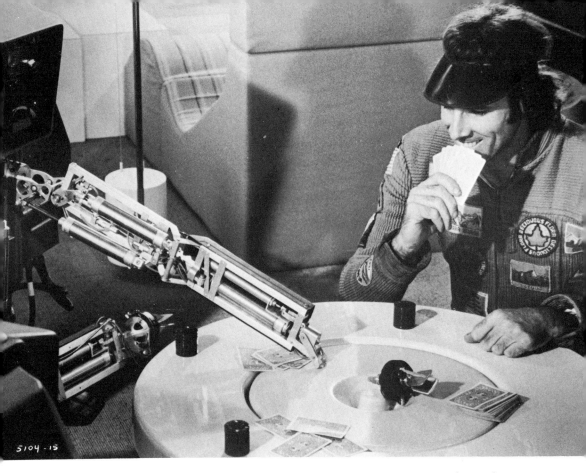

5104-15

The robots of *Silent Running* play a mean game of poker. (*From the motion picture* Silent Running, *Courtesy of Universal Pictures*)

of the crew members—a woman, of course—into a robot and sends her back to the ship to scout around, clad in a white minidress and plastic high heels—what every well-dressed robot wears in space. And as could be expected, Kirk and logic prevail—Earth is saved.

One of the most accurate and interesting portrayals of robotics in the near future is found in the 1972 film *Silent Running*. The plot concerns the attempts of a young ecologist to preserve the last bastion of Terran plant life, an orbiting dome-shaped habitat. He is aided in his quest by three small robotic drones that he has named (after the Disney cartoon ducks) Huey, Dewey and Louey.

These three small triangular automatons march around the habitat on truncated legs, communicating with each other by

The arms of these film robots are not unlike the manipulators of their real-life robotic cousins. (*From the motion picture* Silent Running, *Courtesy of Universal Pictures*)

electronic signals. Their purpose is to maintain the station and its greenery. The ecologist talks to them, orders them, plays poker with them (computers today are quite capable of playing poker, although they do find it somewhat difficult to bluff) and treats them more as companions than as machines.

However, it is clear that it is the character who is anthropomorphizing the robots, not the filmmaker. It is not unreasonable to assume that a person working closely with robots will come to feel an affection for them; this can be observed today in many instances concerning industrial robots. However, it is very unreasonable to expect that a robot, following today's technological probabilities, would form an affection for a human—as occurs in most films dealing with robots.

The dynamic duo: Famous robots R2D2 and C3P0. (© *Lucasfilm, Ltd. (LFL) 1980. All rights reserved.)*

There can be no complete listing of movie robots without mentioning the two automatons that are featured so prominently in the popular space opera *Star Wars.*

The main plotline of the screen saga, which as of this writing includes the films *Star Wars* and *The Empire Strikes Back,* concerns the adventures of the young hero, Luke Skywalker, as he battles a corrupt galactic empire, in the form of the evil Darth Vader. Accompanying him on his quest, along with a motley crew of humans and aliens, are two robots or, in the characters' words, droids. ("Droid" is probably a shortening of the word "android," a term usually associated in science fiction with artificial beings constructed of biological material; however, robots, including those in *Star Wars*, are constructed of mechanical parts. For example, in today's parlance, the "robots" in *R.U.R.* would be referred to as androids.)

The two droids in *Star Wars* are quite different in makeup and personality. C3P0 is the epitome of the intelligent, verbal, humanoid robot so often found in science fiction. It is almost human in its attitudes and reactions—the perfect robotic companion. Its gleaming, smoothly metallic body is highly reminiscent of the female robot in *Metropolis*.

R2D2, on the other hand, while just as intelligent as its partner, is more robotic in design. Balanced precariously on two wheeled legs, the small automaton communicates through a series of clicks and beeps that only its robotic peers can accurately decipher.

In fact, the *Star Wars* universe is filled with a gratifying multitude of robots, most of them mechanical servants to be manipulated by the animate beings around them. A race of desert creatures deals in used robots much as today's salesmen deal in used cars. A robot space probe scans a planet's surface for indications of life. A tiny robotic tool helps repair a space ship, and a robot physician helps repair the hero. There is even a suggestion that the spacecraft are robotic in nature, a'la *2001*, when at one point in *Empire* C3P0 is heard to complain about the atrocious "dialect" of the ship's computer.

The wide popularity of *Star Wars* set off a multitude of film and television imitations. The directors of most of these efforts apparently felt that there were two major elements that were required to make a successful science fiction film—lots of blazing space battles and cute robots. The media was soon inundated with improbable automatons—none having anything in common with the real robots waiting in the wings.

It is neither cute, cuddly nor clever, but it is a real, working, existing robot. (*Jerome McCavitt, Carnegie-Mellon University*)

BEYOND THE DREAM

Films, television series, novels and short stories have all fed our imaginations for over fifty years with images of intelligent, independent robots, but what is the reality behind all these elaborate robotic fantasies? Where are the shining metal legions that have—peopled?—our fictional worlds for so many years? Do they, and can they, really exist?

While writers and directors have been busy creating fictional mechanisms, scientists, technicians, hobbyists and tinkerers have also been busy, in more practical ways. The vanguard of the real robots is already here.

Today's automatons aren't quite as clever, graceful or thrillingly malevolent as those we are used to. In fact, they are comparatively stupid, depending on very detailed programming to perform the simplest of tasks. Rather than resembling the gleaming robotrix of *Metropolis,* they are, for the most part, stumpy machines with long metallic manipulators and no personality whatsoever; or small triangular boxes hesitantly finding their way along a wall; or rather nondescript space vehicles. One's first impulse upon sighting these machines may be to shrug and say, "What's the big deal?"

Only think a moment. A few years ago, these machines did not exist. Now, we have robots assembling delicate radio parts in factories and venturing into nuclear reactors to handle radioactive material. A robot was the first Terran to touch the planet of Mars, and its more sophisticated cousin may soon be strolling across the Martian terrain. Teen-aged students are now guiding simple assemblies through programmed actions that were considered major scientific problems 20 years ago. What will those same students be building 20 years from now?

Nothing is totally impossible. Perhaps one day we will be able to produce robots that are practically indistinguishable from ourselves. In the meantime, it would be wise to examine what we've got now, and where we're going.

And that, in itself, is quite a trip.

We may never see human-shaped, household robots—but far more interesting machines may be in store.

Chapter Two

Real Robots

WHERE *ARE* THE ROBOTS?

FROM ALL THE examples that television and motion pictures have so kindly provided us with, it would be logical to assume that science will be populating the country with intelligent robotic companions any day now. Most magazine articles on robots usually end by telling us that, in spite of what everybody says, intelligent humanlike androids may be just around the corner. And what about that five-foot-high metal man that was giving out free samples at the supermarket last week—wasn't that a robot? Where can you buy one?

Unfortunately, robots as sophisticated and intelligent as those on the silver screen are still things of the imagination. Human-shaped C3P0 had a very human actor inside him; and the few times that little R2D2 was actually played by a robotic machine, instead of a very small man in a robot suit, the machine had an unfortunate tendency to fall over (off-camera, of course). As for the gregarious robot at the supermarket, if you'd made a careful search of the crowd watching its antics, you might have noticed a technician discreetly controlling its activities with a remote device.

Science fiction notwithstanding, there are no robotic nurse-maids minding children or answering the door. You will not find a robot walking down Main Street, U.S.A. (or Main Street, Anywhere, for that matter), tipping its hat to passersby and inquiring the way to the nearest body shop.

Why?

Because they haven't been invented yet.

WHAT IS A ROBOT?

This is certainly not to say that there are *no* robots. It is just that the state of the art of robotics is only in its infancy. While people have been reading and dreaming about mechanical servants since the days of *R.U.R.*, our technology has only recently become capable of fulfilling the first of these dreams. And the paths that the research is following, while beginning to produce incredibly interesting and intelligent machines, is not quite up to producing wise-cracking automatons—not just yet, anyway.

Of course, all of this depends on what you consider a robot to be. As we go further, you will find that most peoples' ideas of what a robot is depend in large measure upon their particular viewpoint. For example, scientists can't agree on whether or not our exploratory spacecraft are, in fact, robots; industrialists differ about what distinguishes a robotic worker from simple automation; and many purveyors of what are called "show robots" (i.e., the supermarket salesman) will insist that there is no reason not to call their machines robots.

Most do agree that, in general, a robot should have at least three basic characteristics: (a) it is mechanical; (b) it is capable of performing certain tasks; and (c) it is equipped with at least some intelligence and/or some independence of action. But here you can run into problems as well.

Take, for example, your normal, everyday washing machine. It is certainly mechanical. It is capable of performing a fairly important task—washing clothes. And, to a certain extent, the machine performs independent actions and makes decisions. It knows when to stop washing the clothing and go into the spin cycle. If too many articles of clothing are piled into it, many machines are even capable of "deciding" that they cannot

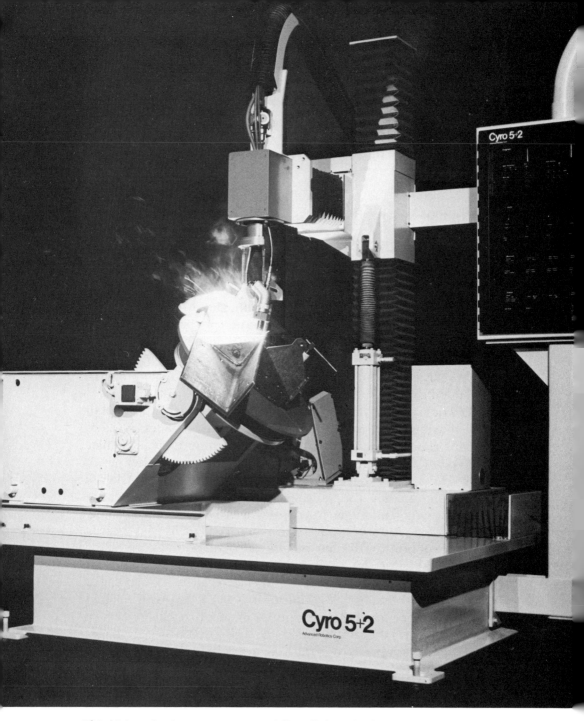

This high-technology automaton satisfies all the robotic requirements: it is mechanical; it's performing its assigned task—welding—quite adeptly; and evidently has enough intelligence to do the job. (*Advanced Robotics Corp.*)

handle the load and will either buzz their dissatisfaction or refuse to run. A robot, right?

Obviously, a washing machine is not a robot (for one thing, it cannot be reprogrammed to hang the wash to dry), although possibly one day there will be a robot that performs those duties. And that robot may well look just like the washing machine.

Aesthetics aside, nowhere is it written that a machine must look humanoid to be called a robot. In fact, the human form would be a distinct disadvantage to most of today's robots. Our arms, our legs, the placement of our eyes and ears and mouth—all are part of a highly balanced, efficient biological assemblage that uses each part to its fullest advantage. Robots, on the other hand, are planned to perform specific, limited jobs for which only a few limbs and abilities are necessary.

An assembly line robot, for example, is often solely made up of its computer plus a huge, jointed, multipurpose "arm" that can be designed for whatever task it's assigned to. Why would it need legs? It's not going anywhere; the work comes to it.

And even when robots get more complicated, they still don't sprout arms or legs or faces. In more complex machines, such as sighted industrial robots, the robot's "eye" is usually located either in the ceiling or in its "hand." Chess-playing computers that are on the market today are shaped like—what else?—chess boards. The Martian rover, a mobile robot that may one day tell us a great deal about the surface of the "red planet," looks like a poorly designed golf cart. Many of the robots built at home by hobbyists look like boxes on wheels, with perhaps a few odd wires straying here and there, giving a clue to the machine's complexity. Only film directors and science fiction writers still have any need to make their robots conform to human ideas of body proportion.

While a robot's outward resemblance to human beings is of no real importance, it must of course be capable of carrying out those commands that it has been built for. Robots are a lot like us in this—they depend on the efficient interworkings of both their physical (or mechanical) and mental (or computerized) systems to be able to function as a whole, useful unit.

Most automatons are built with these two factors in mind: their intelligence and their motor capabilities. The first is supplied by a high-level computer that can be an actual part of the robot, can be connected to it by a cable or through a radio link or can indirectly control the robot by designing programs that are then separately inserted into the machine's system via tape cassettes or discs. The second depends on the robot's function.

THE FIRST ROBOTS

The first robots, like human infants, were rather more of the physical than the mental. "Shakey," built in 1968, was the first complete robot system. Created by scientists at the Stanford Research Institute in California, Shakey closely resembled an air conditioner on wheels, and was capable of avoiding obstacles, selecting specified boxes and pushing them together in a certain area. While at first it was connected to its controlling computer by a cable, it was later able to perform the same functions using a radio link.

Another landmark robot was built several years ago at the Johns Hopkins Applied Physics Laboratory. Scientists there were attempting to create a totally independent robot, and one of the first criteria for such a machine was that it should be able to "feed" itself.

The robot they developed looked like a metal hatbox and was powered by a rechargable battery. It was programmed to explore its environment, and that's what it did—feeling its way around the laboratories at Johns Hopkins using a metal arm and ground sensors that prevented it from falling down the staircase. Eventually the battery would begin to get low, and this would trigger a preset detector. Using its arm, the robot would feel its way along the nearest wall until it felt an irregularity—a wall socket. It would then plug itself into the socket, using prongs in its arm, replenish the battery and, when its "hunger" was sated, unplug itself and continue on its travels.

Coming between these two early robots, and following them, were a multitude of increasingly sophisticated automatons. It is

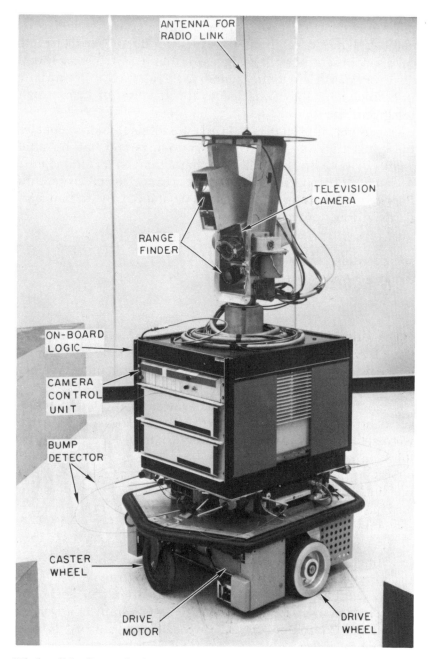

ANTENNA FOR
RADIO LINK

TELEVISION
CAMERA

RANGE
FINDER

ON-BOARD
LOGIC

CAMERA
CONTROL
UNIT

BUMP
DETECTOR

CASTER
WHEEL

DRIVE
MOTOR

DRIVE
WHEEL

"Shakey," the first real robot. Using its television camera and bumpers, Shakey could locate objects and reassemble them in set patterns. (*SRI International*) difficult to chart the exact course of advances in the field, or to name each new development, because for the most part technical development doesn't really work that way. While each new discovery may seem to the layman to be relatively unimportant, in combination they have led to the creation of increasingly intelligent robots.

One important fact to note is that the Johns Hopkins robot felt no actual curiosity about its surroundings and had no intellectual drive to find out what was happening in its limited environment. It had been ordered by its creators to roll around the floor of the laboratories, and that's what it did. By human standards, it was not terribly smart.

HOW ROBOTS THINK

The study of the way computers and, by extension, robots think is called artificial intelligence (AI) or machine intelligence. (Since nobody has ever satisfactorily defined what intelligence is, some scientists object to the term "artificial.") Interestingly enough, the study of AI originally began as an attempt to understand how the human brain worked. Electronic schematics were set up to try to create elementary models of the human thinking process. It didn't work—students of the brain have gone on to other methods of investigation. But somewhere in the building of those first simple problem-solving computers, the emphasis shifted from finding out how *we* think to trying to make the *computer* think.

To a large extent they succeeded. Computers today can play an irritatingly good game of chess and can run robots through programs involving difficult and precise movements. Industrial robots can efficiently weld along specified lines. Spacebound craft can make soft landings on alien planets almost entirely on their own. Computers have even been programmed to play psychiatrist and schizophrenic patient so well that doctors have had trouble distinguishing accounts of their conversations from those of human doctor/patient exchanges.

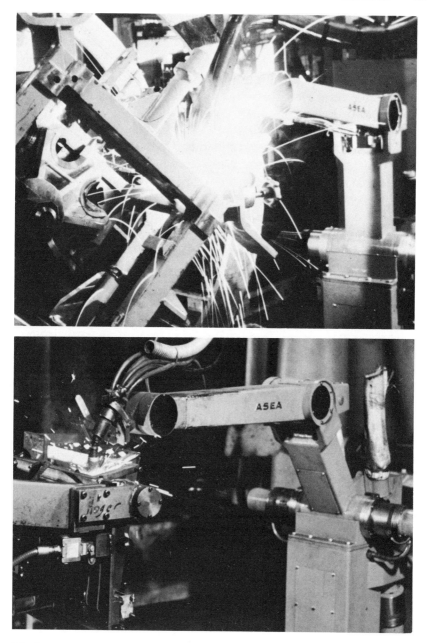

The best thing about the way robots think is that they don't get bored. These workers will do this dirty, dangerous welding job hundreds of times a day and never even consider goofing off. (*ASEA*)

But there are limits to their thinking processes. Chess has definite rules, a very precise structure. A computerized robot can be told that if its opponent makes move A, and the other pieces are in thus and such positions, then move B is the most desirable. Complicated, sure—but all following set mathematical and thoroughly programmable rules.

A machine's programming is the set of instructions that tells it all about the world and how to respond to it. It may seem easy to tell a machine how to do anything you do—but that's only on the surface. Try something simple—write down *exactly* how you tie your shoelaces, complete with what to do if the lace is wet or muddy or already in a knot—explain it so that someone who has never seen a shoe, much less a shoelace, can understand. Now, do that with everything else you do in a single day. Seem even the slightest bit impossible?

Carl Ruoff is supervisor of the robotics and teleoperatives group at the Jet Propulsion Lab in California, and is very much aware of the intellectual shortcomings of robots today. "A lot of the robots that are used today," he says, "really have very limited ability to determine whether or not they've accomplished their task, and very limited ability to adapt themselves to variations of their task.

"Now, that isn't always bad. You could regard the industrial revolution as creating automatic devices that could do tasks people don't choose to do. Most of these machines do not have much ability to adapt. But the humans stay near them and do the adapting for them.

"A lot of things you buy, from Chevrolets to canned peaches, are all put together by various kinds of machines, simply because people couldn't do it economically. In the past, most machines had to be adjusted constantly by people. For example, if you have an automatic canning machine, and the cans aren't coming in at precisely the right position, then instead of sealing them when the plunger comes down, it would just smash the can. That's because the canning machine is not able to sense where the can really is. Now, if you made the canning machine smarter, perhaps gave it some computer vision or other sensors, then it could be made to adapt to where the can really was."

Robots have reached certain levels of understanding. Even less complex robots can memorize certain movements that they are "taught" by human technicians, and follow those movements faithfully. A few are now aware to the point that, when a mistake has been made, they can take limited corrective action and/or notify the operator that something is wrong.

But they cannot adapt to new situations, and since almost *everything* is a new situation to a robot, this can create problems. "Computers can generate very high-level plans like 'pick this up and put it down,' " says Ruoff. "But something like 'pick this up and put it down' doesn't really mean anything unless you've defined what 'this' is and where 'down' goes—established a real universe for a real problem.

"Say you're going to pick up a pencil. Maybe the pencil is the wrong color, or it's in a little bit of an awkward position, or there's a rattlesnake near it so you have to pick it up carefully." He smiles. "You may be able to cope, but all these things conspire to make artificial intelligence programs really screw up! It's a problem of the variability of the world."

However, all this still doesn't explain why we cannot build intelligent mobile robots. Why not simply program a knowledge of the world into the machine and set it loose?

Well, aside from the huge programming task that that sort of approach would entail, the outside world is not a robot's only problem. Artificial intelligence encompasses not only what a robot *thinks*; it also has a great deal to do with how a robot *moves*. Making a computerized machine that can do the things a human being can do would be very simple—if human beings were simple machines, which we certainly are not.

Carl Ruoff puts it this way: Imagine that you are seated and have decided to stand. You first look around to see that there aren't any obstacles in the way. Meanwhile, you are unconsciously performing such acts as monitoring your balance, shifting your weight from your seat to your legs and feet, perhaps taking a stronger grip on the armrests of the chair in preparation for the act of standing.

As you stand, you are feeling the floor with your feet, adjusting your knees and legs to rebalance the weight, letting go of the chair arms at the correct moment (after all, you don't want to take the chair with you) and then bringing your hands forward to aid in balance. All this is done without really thinking about it.

A robot would have to think about it—every single step of the way. It would need the ability to see. It would need tactile (touch) sensors to tell it when its feet were planted firmly on the floor. It would need proper programming to help it shift and maintain its balance. It would need to know at which exact moment to release the chair and where to put its hands to maintain optimum balance.

It would also need a very sophisticated program to recover from any mistakes it made in this seemingly simple move. Now, one of a human being's most useful talents is the ability to recognize when s/he has done something wrong, stop, think it over and then proceed to take the correct action. Say that as you are preparing to stand you notice that one of your legs has fallen asleep. You instinctively know enough to immediately sit down again and wait until there is enough circulation in the leg so that it will support your weight. However, if something went wrong with the leg joint of our standing robot, it would need an appropriate error-recovery program to tell it that it is presently incapable of standing, and that it should sit back down. Otherwise, it would either continue standing, and fall over, or stop in total confusion, and fall over. Either way, the result is not desirable.

All the preceding descriptions just concerned a very simple physical movement: the act of standing. For us, it's an easy process, but for a robot, it's a complicated problem. Not impossible—just difficult and, today, more costly in time and effort than it's really worth. "You can try to program the robot by putting tactile sensors in it," explains Ruoff, "and program all those tactile sensors explicitly to recognize what an object felt like or to recognize that it felt appropriate. It starts to be a very hard programming task that can take just months and months and months, and is very difficult to debug [find and correct the problems.]"

VISION

One of the prime sensory abilities that scientists have been working very hard to perfect is sight. Vision is an extremely useful tool to have (especially for a machine that has no sense of touch to begin with and relatively little intelligence), and so it is high on the list of developmental priorities.

If you're talking about an industrial robot, then the machine that can see will be much more able to correct errors than one that cannot. Instead of reaching for a nail and closing its hand-piece on air (and then blithely trying to drive the air into a piece of wood), a robot that can see will be able to look at the nail and not only verify that it's there, but also that it's the correct size and in the correct position, and being driven into the correct piece of wood.

Vision also makes it a lot easier to program a mechanism. Say you had a robot whose job it was to place a record on a stereo turntable. If the turntable was exactly where it was supposed to be, and the record was exactly where *it* was supposed to be, a nonsighted robot would do fine. But if the turntable had been pushed slightly askew, or the record grasped a little off-center, the robot wouldn't know (or care), would simply place the record exactly where it was programmed to and would probably break it in the process.

If that same robot had been able to see, it would only have to be told that the record will be approximately *here* and the turntable approximately *there,* and it could verify for itself the exact position of the equipment. Less programming would be needed to ensure that the robot worked properly (there are fewer steps involved in telling a computer "approximately" than in telling it an exact location, point-for-point), and you wouldn't lose as many records.

Intelligent vision capabilities would also be incredibly handy to have on board any of our unmanned spacecraft. The Martian rover may have two different types of vision systems working to allow it to keep out of ditches and track courses to interesting sites: a laser system and a three-dimensional camera system.

So what's the big deal, you may ask at this point. Simply hook the thing up to a television camera and tell it what to look for.

Once again, it isn't quite that easy.

For one thing, like our "standing" example, a robot does not see the way a human does. When we view a scene, we see it as a whole, meaningful image. But there is nothing in a robot's experience to make that same image meaningful. In fact, its memory is not complex enough to be able to simply "see" a total picture. Since a computer does not speak the language of light the same way our brain does, the image therefore must be translated into a language the robot can understand: the language of mathematics.

Vastly simplified, the process goes something like this: Suppose the robot is told to look for a bright silver quarter located in the center of a black shelf. The image is first recorded by a solid state television camera. What the camera is doing (and what any television camera does) is scanning that scene as a series of slices as the camera's electron beam swiftly goes from left to right, drops down a line and scans left to right again. (Look closely at your television screen and you'll see those slices quite clearly.)

But that is still too complex for the robot's computer to understand. It has to divide the picture further, into a series of dots, much as a printed photograph is composed. This gives the computer a specific map to work from—for example, it can locate the third dot from the left on the 18th line down, if it is asked. The dots that make up the computer's field of vision are called *pixels*—shorthand for *picture elements.*

Now the computer can locate anything it sees by converting the scene into pixels. A human looking for that quarter would probably try to find the relative brightness of the coin against the dark background of the shelf. The robot essentially does the same thing—except, unacquainted with such concepts as dark and light, it needs a more mathematical criterion to follow.

As the television camera picks up the image, it is converting the light waves into voltage peaks—in other words, the voltage of the signal that the computer is getting goes up as the light becomes brighter, and goes down as the light becomes dimmer. A digitizer—literally, a computer that converts signals into numbers—takes the varying voltage peaks and gives each a numerical value. So that while a totally dark surface would rate a zero (since there is no voltage to speak of), a moderately bright surface would rate 100, and an extremely bright surface would rate 255 (which for many computers is the highest numerical value assigned).

Let's backtrack a moment. The robot's television camera is picking up the image of the quarter on the shelf, converting that image into pixels, and rating each pixel with a numerical value according to its brightness. Now the computer has a picture of that quarter in values of, say, 150 on a field of tens (the brightness of the shelf).

Now, very few images are made up of only two levels of brightness. Even with so simple a thing as a coin, there are shadows and highlights to consider. The next step in the process is to tell the robot's computer the approximate brightness of the quarter it will be looking for. This is called *threshholding*—the programmer is giving the robot a certain threshhold value; any pixels that fall below that brightness value will now be given a new value of zero, while everything as bright or brighter will be given a one. So you've got a group of ones representing the quarter—a digital picture that looks something like this:

```
0  0  0  0  0  0  0  0  0  0  0  0  0
0  0  0  0  0  0  0  0  0  0  0  0  0
0  0  0  0  0  1  1  1  0  0  0  0  0
0  0  0  0  1  1  1  1  1  0  0  0  0
0  0  0  0  1  1  1  1  1  0  0  0  0
0  0  0  0  1  1  1  1  1  0  0  0  0
0  0  0  0  1  1  1  1  1  0  0  0  0
0  0  0  0  0  1  1  1  0  0  0  0  0
0  0  0  0  0  0  0  0  0  0  0  0  0
0  0  0  0  0  0  0  0  0  0  0  0  0
```

Now that the image of the coin has been translated into terms that the robot can understand, the computer will (electronically) tell the robotic arm the exact location of the quarter and which movements it must go through in order to pick it up.

Again, this is a very basic recreation of computerized sight. There are other, more complex and efficient methods; in addition, there are other ways to use this same method. For example, some factories will place their parts on a translucent table that is lit from underneath. This puts those parts into silhouette; the computer is instructed to look for pixels that come *under* a certain value. Additionally, the computers must be instructed how to overlook such unnecessary objects as motes of dust that may also fall into the appropriate brightness category, or how to distinguish between two similarly shaped parts.

And they can do it! There are robots in industry today that are becoming more adept at their jobs because of their new vision capabilities.

One of the first uses of robotic vision, and one that has become quite common today, is the employment of television cameras to look for serious flaws in parts passing on an assembly line. When the machine's programming signals that something is wrong, an airjet blows the piece off the assembly line and into a reject bin.

Something a bit more complex is an assembly robot developed by the Stanford Research Institute. The automaton, which uses its vision more directly than the airjet robot, is essentially made up of three separate components: an industrial robotic manipulator or arm (in this case, one made by Unimation, the first corporation to solely produce robots), a television camera in the ceiling and a programmable, moveable table on which there are all the parts waiting to be assembled. The procedure goes something like this: The television camera photographs the part that is to be worked on (in this case, an automotive compressor), finds its coordinates and orders the table to move so that the compressor will be at the center of the camera's point of view. The robotic arm can now place a cover on top of the compressor.

When that is done, the computer again notes the compressor's position and again orders the table to move, this time so that a hole in the rim of the cover is central. The arm can now place a bolt through the hole and attach the cover to the compressor. The same procedure is followed for several other holes in the rim. The fascinating thing about this robot is that, in this case, the robot is not only made up of the manipulator arm and

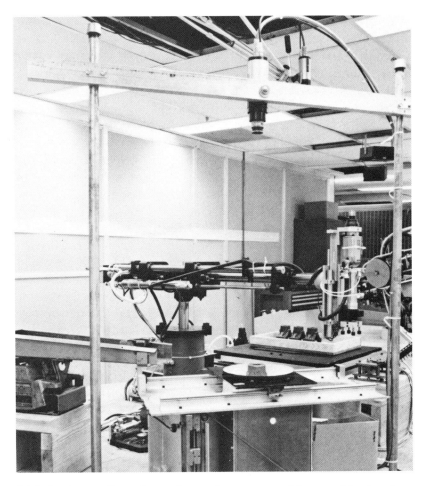

This robot is actually made up of several separate machines working together: the camera which locates the object to be assembled (above), the table which moves the object into position (front, center), the arm which does the actual assembling (back, center) and the computer which gives these machines their orders. (*SRI International*)

its computer, but also the camera viewing the scene and the table that obediently obeys the computer's orders.

This is only the beginning. In spite of all the advances made (most of them within the last few years), scientists are trying to bring robotic vision still closer to that of its human counterparts. For example, while robots can now distinguish objects and shapes with a great deal of success, technicians have only begun to solve the "nit-picking problem"—that is, enabling a robot to choose a needed object out of a group of unrelated, randomly placed objects (for example, picking a hammer out of a box of tools). And when it comes to recognizing more abstract images, robots are almost totally incompetent.

"You take a computer," says Carl Ruoff, "and try to get it to recognize scenes, or recognize a human. You know how you can see a picture of your mother and instantly, in a few milliseconds, you recognize her? It would take a computer hours—and if you had a bunch of similar-looking people, I don't think a computer could ever solve the problem of differentiating your mother from other people. It *might* be able to say 'This is probably a woman because she has long hair'—but if a woman happened to wear short hair, it might think she was a man."

SO, THAT'S THE PROBLEM

All this sounds a bit discouraging—and the above problems are just the beginning. There are many factors to be considered when planning and building an efficient, able, intelligent robot, such as circuitry, programming, balance . . . in fact, it would take an entire book just to detail the structure and circuitry of a single, simple robot—and there are several such books available (some of which are listed in the bibliography). This is not one of them.

You might want to consider this chapter as sort of a preparation—it is hard to be impressed with today's robots if you are thinking in terms of their fictional predecessors. Yes, there are many difficulties involved in developing a sophisticated robotic system, and no, there are no humanoid robots in our *immediate* future. In fact, we haven't even fulfilled some of the short-term dreams we thought we would have fulfilled. "People said, years ago, that in a few years it should be possible to have cars riding around automatically with computer vision," recalls Carl Ruoff. "And we're not anywhere near being able to do that in any kind of robust, sophisticated way. One person said that it has taken 18 years to get as far in artificial intelligence as he thought they would be able to go in about ten months."

But while robots remain something of a puzzle, it is a matter of record that most humans find nothing more challenging than a puzzle—especially one that everyone said could not be solved. It is also a matter of record that people *want* robots—as workers, as extensions of themselves and as companions.

And, as many of the people quoted in the coming chapters will testify, they are getting them, but not in exactly the way pictured by writers and film directors. That makes it even more interesting, doesn't it?

Chapter Three

Blue Collar Robots

TODAY'S INDUSTRIAL ROBOTS probably make up the largest robot population in existence today. Unfortunately for the eager science fiction reader, on first examination they are a rather unromantic bunch.

For one thing, they are almost totally unrecognizable *as* robots — to the uninitiated. Walk into any factory in which robots are engaged in, say, screwing bolts onto a piece of metal, and you might simply shrug at yet another example of mindless automation. The working automatons, dipping and raising their limbs like a line of metallic Rockettes, do not have voices or interesting blinking lights; many of them cannot even move away from their stations, and they certainly do not resemble anything even vaguely human.

In addition, they are annoyingly single-minded. Unlike the "show robots" you may meet in shopping centers, an industrial robot will totally ignore any human it may come into contact with (except its programmer). In fact, if that same human is injudicious enough to get in the way of a robot intent on its job, it would be just as happy to weld a door onto him as it would to a car.

But while a science fiction fan may be disappointed in these machines, most manufacturers will agree that, while neither interesting nor romantic nor particularly smart, industrial robots are slowly revolutionizing the workplace. And while the state of today's robotic art is still comparatively primitive, it is improving at an increasingly rapid rate.

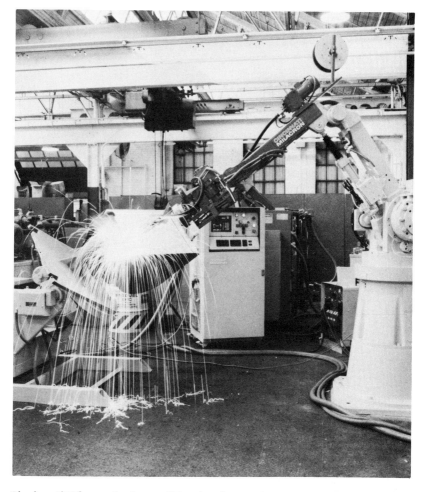

Cincinnati Milacron's advanced T3 robot, here using an attachment to weld two parts together. Robots such as the T3 are capable of doing highly specialized factory work, 24 hours a day, 7 days a week—with no coffee breaks. (*Cincinnati Milacron, Inc.*)

There are some 3500 robots hard at work in the United States today. This is a 35 percent jump from last year's figures, and experts are counting on a further 35 percent increase during the next five years. They can be found in large numbers at such automobile manufacturers as General Motors, Ford, Chrysler, Honda and Fiat; also at such companies as Westinghouse, General Electric and Texas Instruments. They are performing a wide

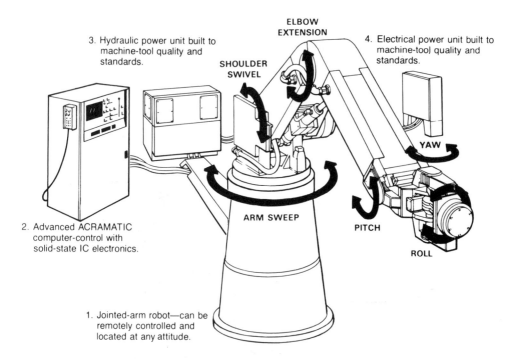

3. Hydraulic power unit built to machine-tool quality and standards.

ELBOW EXTENSION

SHOULDER SWIVEL

4. Electrical power unit built to machine-tool quality and standards.

YAW

2. Advanced ACRAMATIC computer-control with solid-state IC electronics.

ARM SWEEP

PITCH

ROLL

1. Jointed-arm robot—can be remotely controlled and located at any attitude.

A diagram of the T3 robot. On the left is the robot's computer—the ''brain'' of the operation. The power source is next to it. The robot itself is built so that both the arm and the manipulator can move in a variety of directions: side to side (''yaw''), up and down (''pitch'') and in a circular motion (''roll''). There is a tendency, even with such a nonhuman-looking machine, to compare it to the human form—notice the labeled ''arm,'' ''shoulder,'' and ''elbow.'' (*Cincinnati Milacron, Inc.*)

range of tasks, from the simple movement of heavy equipment from one area to another, through the painting and welding of car bodies, to the delicate assembly of radios and electronic parts. They are highly sophisticated, useful, functional machines.

Therefore, the appearance of these automatons does not trouble manufacturers in the least. Lawrence Kamm is president of the Modular Machine Company, which manufactures very simple, inexpensive robots that do uncomplicated tasks such as loading and unloading, or sorting parts. "Many of the people in the robot business are trying to make a robot with human characteristics—an imitation man or Frankenstein," said Kamm in a newspaper interview. "But that's the last thing a manufacturer needs. He wants a machine that is economical and makes products. He doesn't want entertainment toys."

The imposing metal creatures that grace our automobile factories could hardly be called entertainment toys. Many of those used to handle heavy equipment stand taller than six feet and weigh in at about 1500 pounds—and are able to pick up a 60-pound part with no difficulty.

The typical industrial robot—if there can be said to be a "typical" robot—consists of a large, rectangular base to which is attached a long, hinged, multi-functional arm or manipulator. Extending from this arm is the robot's "hand," the exact specifications of which vary according to the machine's function. For example, it can be a claw-like gripper for picking up parts, a punching or screwing mechanism, or even a spray nozzle.

Since the first industrial robots were introduced in the early 1960s, they have mainly been used for jobs that human workers can not or will not carry out. This is hardly surprising since the typical robot can cost anywhere from $10,000 to $150,000 (depending on its abilities and sophistication). Most corporations can hardly afford to purchase one just for the fun of it.

HAZARDOUS DUTY

Very often, the industrial robot will be assigned to tasks too hazardous for our relatively fragile frames. One of the more well-publicized examples of life-saving robotics is Herman, a remote-controlled mobile manipulator designed by the nuclear division of the Union Carbide Corporation. Strangely enough, by

"Herman," the robot which stood at ready during the Three Mile Island nuclear plant crisis. (*Union Carbide*)

most definitions now in use Herman cannot really be called a robot. It is in constant communication with both its computer and a human controller who operates it through a 700-foot-long "umbilical cord."

But Herman is nevertheless a very good example of the importance of robotics in today's high technology world. During the March 1979 accident at the Three Mile Island nuclear plant, Herman was standing by in case it was needed to venture into

Herman is actually made up of several separate units. From left to right: the mobile manipulator, a remote control arm assembly, the power unit and the operator's console. Since Herman is under constant human supervision, he can really not be considered a truly independent robot. (*Union Carbide*)

areas of high radioactivity—areas that would be harmful to even a shielded human technician.

Nuclear plants are not the only industries where hazardous jobs are being assigned to robotic workers. There are plenty of manufacturing facilities around the country where the inherent dangers of the machinery have made employers all too happy to hire robots.

Joseph T. Engelberger is founder and president of Unimation, Inc., one of the oldest and largest robot manufacturing firms in the world. "You might have a man holding up a 60-pound welding gun," explains Mr. Engelberger. "He's got to move it around and take all the flanges of the car body and weld them together. A few hours of that is very tiring. That's probably become the number one job that robots do now.

Robots have become commonplace in most automobile factories, where they are stationed along moving assembly lines, doing their jobs with unvarying, untiring, unthinking regularity. (*Unimation, Inc.*)

"We also do a lot of work molding plastic. You know those big plastic garbage cans? It's a very dangerous thing to have a human walk between the dies [molds]. There have been fatalities. If a robot happens to leave an arm in there, there's no harm done — just put another arm on the robot."

A robot can take a red-hot casting from the mold in which it was formed, lubricate the mold, quench the steaming casting in water and place it in a machine that will trim the edges. A robot can pick up an awkward kitchen sink and guide it easily against buffing wheels to bring it to an even shine. A robot can spray-paint a car or a wooden chair, following exactly the right move-

A robot about to plunge a steaming hot part into cooling water. Robots were first used on such dangerous jobs as die casting. (*Unimation, Inc.*)

ments to thoroughly and evenly coat the object with paint, without worrying about inhaling the dangerous fumes. And these robots can do each of these jobs over and over again, exactly the same way every single time. People can also do these jobs, but fatigue or exposure to dangerous conditions or chemicals creates a risk of injury to a worker who's simply so tired that he or she can't do the job correctly after a little time on the line.

THIS IS A ROBOT?

What makes these machines robots? Are they not much more than technically improved examples of the type of automation that has been in use since the assembly line was invented?

To answer that question, it becomes once again necessary to try and define what a robot is. Mr. Engelberger, who is in demand as a lecturer on industrial robotics, likes to show as part of his presentations a set of four slides, each bearing four different definitions of robots. The first is derived from *Webster's Seventh Collegiate Dictionary,* which defines a robot as being either a device acting in an independent way, or a human acting like a robot (a useful, if rather roundabout statement).

The second, from industrial sources, states that a robot is a "multi-articulated manipulator which can be programmed to carry out a variety of different industrial tasks."

By the third slide, things are getting a little more informal. Constantly asked by the press for his definition, Mr. Engelberger has come up with an unarguable one: "I can't define robots, but I know one when I see one.

"The final thing I say," he relates, "is that the Robot Institute of America had a committee to come up with a definition and this definition got to be a page and a half long. They gave up in dismay, and concluded that the only definition for a robot was 'If you pay your dues to the RIA, you're producing a robot.' "

Joking aside, the very fact that there *is* a Robot Institute of America says something about the growing use of robots in industry today. An offshoot of the Society of Manufacturing Engineers, the Institute was formed in 1974 as a response to the growing interest in robotics in the industrial community. Boasting 15 robot manufacturing members, and 55 user members, the RIA tries to smooth the entry of robots into the industrial marketplace.

However, the question of what makes these machines robots still remains. It might be helpful to take a look at the real definition that the Robot Institute of America finally came up with in November of 1979. It reads:

"A Robot is a *reprogrammable* multi-functional manipulator designed to move material, parts, tools, or specialized devices through variable programmed motions for the performance of a *variety* of tasks." (Emphasis mine)

Actually, the emphasis is not totally mine. Don Vincent, manager of the RIA, stresses these two points as most important in determining that a robot is a robot. According to the above statement, a true robot does not simply perform a task. It will not simply solder a wire or hammer a nail. It can be programmed to *either* solder that wire or hammer that nail—and later can be reprogrammed to do something else.

Now, if all these distinctions seem rather mundane and pointless (after all, what's the difference between a machine that will perform one function and one that will perform two?), there are those to whom they are very important indeed—those to whom a slight lead in the industrial sweepstakes could mean success or failure.

THE COMPETITIVE EDGE

Fluctuating global economies have caused the corporations of various major industrial nations to engage in what is politely known as cutthroat competition. Two of the main contestants are the United States and Japan, especially over automobiles and electronic components—two of the most roboticized industries.

At first glance, Japan seems to be winning the robot race. With its well-known enthusiasm for technical advances, and because of wide acceptance of robots among both management and labor, Japan claims to have taken the lead both in robot manufacturing and employment. In 1977, for example, there were about 120 robot manufacturers in Japan as compared to 30 to 40 in the United States and Europe.

American robot manufacturers, however, disagree with these figures, and say that it is all a matter of what one *says* is a robot.

According to Don Vincent, "The Japanese say there are 40,000 [robots] in their country; but by our definition, the Japanese probably have about 12,000. What they call a robot is any automatic device that does something. Here, we make it a little more specific—it has to be programmable.''

"They have absolutely no technical advantage over the United States," agrees Engelberger, "but they sure have an advantage in dedication and expenditures. There's a commitment. The users are buying in large quantities, they're committed to the technology, they get better acceptance from management, better acceptance from labor, they get better government support." But Mr. Engelberger also sees a brighter side to the competition. "All this publicity that the Japanese are getting is making people in the United States and Europe sit up and take notice."

Apparently, one problem has been the reluctance of United States corporate managements to accept the idea of installing robots. This wasn't without reason: until recently, automatons were simply not useful enough to justify replacing human workers in any but the most difficult or dangerous jobs. The cost of the machines, which had to include adjusting the workplace to accommodate the robots, was prohibitive to say the least.

But things are changing. Costs have started to come down (electronics being perhaps the only industry today free from the ravages of inflation—they seem to be getting cheaper while everything else goes up). More importantly, the types of robots available, the jobs they are able to do, and the ease with which a typical non-technician can use them, are making these intelligent machines seem a great deal more practical to a great many more people.

FIRST WAVE

In fact, the use of robotics in industry today has reached the point at which users are beginning not just to talk about "robots" as a single entity—they now discuss the relative virtues of low, medium and high technology robots. The lines drawn between these three classifications tend to be rather fuzzy (how can you classify what you can't define?), but it all has to do with the sophistication of the robot's computer, the number of movements the machine can manage and the variety of jobs it can be persuaded to perform.

Perhaps the simplest way to describe the differences between high, medium and low technology robots would be to compare their actions to that of a human. (Incidentally, there is a strong tendency among robot employers and manufacturers to anthropomorphize their metallic workers. Why, for example, is Herman called Herman? Why not simply the Y-12 mobile manipulator—Y-12 for short? You will also find otherwise hardened engineers talking about the "wrists" and "brains" of their creations. . . .)

Don Vincent has used the example of a person in the act of picking up a glass of, say, water. A low-technology robot (such as the machine produced by Unimation, Inc. that you can see on page 47) would just be capable of picking up that glass and putting it down in a predetermined place. These machines are also known as pick and place robots—they cannot do much more than pick up an object and place it somewhere. These otherwise innocent automatons are the main cause of the confusion as to how many robots there are in existence—since many robotists assert that they are not really robots at all.

There is no problem calling a medium-technology robot a robot. This widely used machine (most robots in use today in factories are medium technology types) would be able to do quite a bit more with the hypothetical glass of water: pick up the glass, turn it on a 180°arc and empty it, turn it back along that same arc to an upright position and then either replace the glass in its original position or set it somewhere else.

The line begins to blur when you try to distinguish a medium- from a high-technology robot, such as Unimation's new PUMA model (see p. 54). The latter might have the glass refilled before setting it down; might be capable of turning the glass in a complete 360°circle; might be able to tell from the weight of the glass whether it was completely full; or could handle a more fragile glass than its less sophisticated cousin.

Once again, some of these distinctions may seem so small as to make no real difference—hardly worth the extra hundreds or

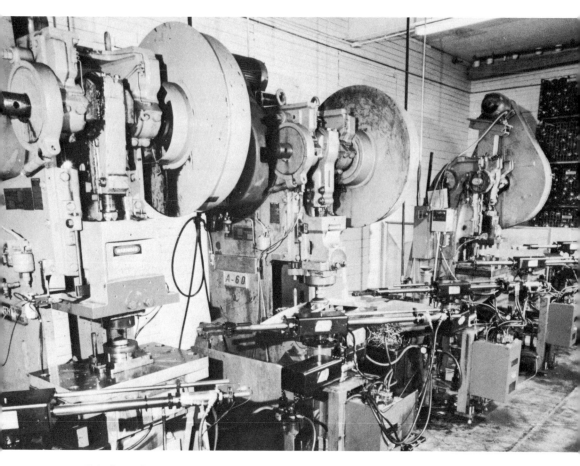

This five-robot system handles the metal stamping of parts for use in the agricultural industry. Only one human material handler is now needed to run the entire assembly line. (*Auto-Place, Inc.*)

thousands of dollars that the more advanced robots cost. But to manufacturers who have to think of the delicacy and exact specifications that go into their products, the cost may be well worthwhile.

Phillipe Villers has described these high technology robots as the precursors of a new era in robotics. Mr. Villers is the founder and president of Automatix, a new robot supply company that has only been incorporated since January 1980. He hopes that his new company will be able to stay afloat in the very unstable

Unimate's new PUMA robot. Notice the difference in size between this highly sophisticated unit and its more clumsy brothers. (*Unimation, Inc.*)

robotics industry market by being the first of a new type of robot "store": the systems supplier.

"We wish to become the leader in a field that has not hitherto existed," explains Villers, "which is robotics systems. That is, taking the robots, the computers, the software, the advanced senses such as vision, and putting them all together for the customer, a function that users presently have to perform for themselves."

SECOND WAVE

It is this type of modular robot that is the beginning of what Villers terms a new wave of robots. "Basically," he says, "the first wave of robotics is essentially powerful material movers, dumb brutes carrying heavy loads in very adverse environments. Second wave robots, which are really just arriving, are generally lightweight, nimble and intelligent. Intelligent in two ways: first, they can adapt themselves in certain ways to their environment because of advanced senses such as vision. Imagine the difference between working blind or sighted in almost any industrial environment—there's a tremendous difference. So vision is the most important of the advanced senses and many of the new robots have it.

"The other increase in robotic intelligence is computing power that's able to make use of what the senses say, to adapt to whatever the task is. And that's basically what the second wave robots are: intelligent robots with advanced sensors and more intelligent computers."

Two of the best publicized members of Mr. Villers' "second wave" are Unimation's PUMA and Cincinnati Milacron's T3 robots. While these two automatons have been built for vastly different types of work, they are both precursors of a more intelligent and sophisticated robotic entity.

PUMA is an acronym standing for Programmable Universal Machine for Assembly. Developed by Unimation and General Motors, PUMA is a smaller robot with the programmability and computer intelligence necessary to assemble delicate machinery; it is therefore one of the smallest and lightest robots on the market today. For example, the PUMA series 250 has an arm weight of 15 pounds, with its accompanying computer adding only about 75 pounds. This makes a grand total of some 90 + pounds—less than a medium-weight human. Compare this with the weight of the usual assembly robot (the huge machines that handle, say, welding chores), that can tip the scales at approximately 1500 pounds, and you begin to see the advantages that the PUMA offers.

"The PUMA was derived from design concepts we had and from research that General Motors did," Engelberger explains. "Ninety percent of the parts in an automobile weigh less than five pounds—which is kind of a surprise to me; that there are all these 3000 pound behemoths on the road and there are so many of these parts, and 90 percent weigh less than five pounds.

"So now we had a standard. We wanted a machine that could

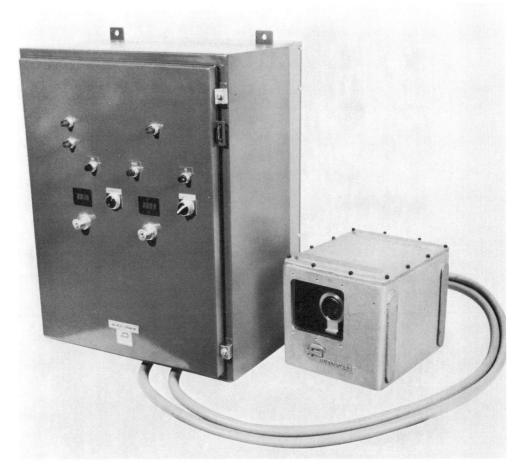

These are the basic units used for a computer-controlled vision system created for Auto-Place robots. From left to right, it is comprised of a control cabinet and a solid-state image camera. (*Auto-Place, Inc.*)

stand in an assembly line in a space occupied by a human, and could work in an indexing line currently used by human workers, and accomplish that work at approximately the same speed. So PUMA was born.''

The robot, which is handling such delicate work as placing light bulbs in sockets, will probably prove quite popular for a number of tasks unsuited to the larger robotic units. In fact, PUMA can be programmed for such fragile work that one of Great Britain's largest candy manufacturers has considered buy-

The PUMA is capable of performing very precise, exacting tasks. (*Unimation, Inc.*)

ing 160 PUMAs to place pieces of chocolate candy in boxes—a job, it must be admitted, that would take a good deal of care (and a machine with no sweet tooth).

PUMA can be programmed in several ways. It can be taught through the playback method, which, as the name suggests, means that the operator guides the robot's arm through the motions it will be expected to follow. The robot "remembers" these motions, incorporates them into its programming and follows them exactly when ordered to. (This type of programming is commonly used with the more artistic painting robots that can be taught to paint anything from the side of a barn to graffiti simply by being guided through the motions once.) The PUMA can also be given orders by inserting the correct programs into the accompanying computer system.

This machine offers multiple advantages to the user. One of the largest expenses for manufacturers is retailoring their assembly lines or working spaces to fit around one of the mechanical behemoths. Plus, the ease of programming and reprogramming these small machines saves both time and money for the operators.

The small PUMA robot is distinguished from its peers by its versatility and relatively low cost—and these have apparently proved most attractive. "We thought we were going to build ten of these things," says a bemused Engelberger. "We have 140 of them on order, half of them by people who never even *saw* it."

Cincinnati Milacron's T3 ("The Tomorrow Tool") is something of a robotic breakthrough. A highly sophisticated machine, it is essentially not all that different from the other robots now being utilized by industry—it's just better.

One of the biggest advantages of this particular robot is that it is easily programmed by employers and/or employees with a minimum of technical training. A small, hand-held box about the size of a television remote control is used to train the T3. As the operator moves the robot through its desired paces, s/he pushes the required buttons on the unit to insert the moves into the robot's memory. Once it stores all the movements that will be required of it, however complex, the T3 is ready to go to work.

T3 is also a very versatile manufacturing tool. Unlike most robots that can only do involved work on motionless objects, the T3 can weld car bodies on a constantly moving assembly line. It is also capable of doing several different jobs in sequence, rather than being limited to repeating the same task over and over, because it has a more intelligent, flexible computer. Also, its manipulator "arm" has a wider range of movement than others of its type.

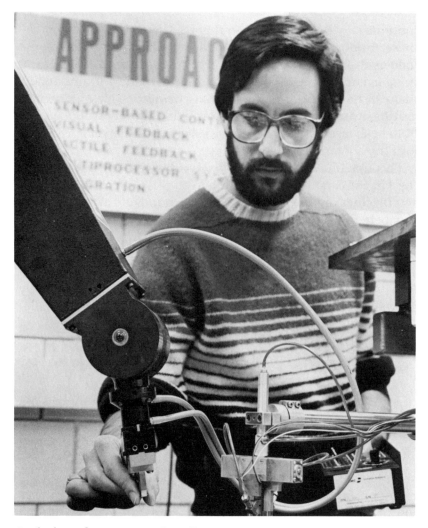

A robotist at the new Carnegie-Mellon Institute for Robotics checks out a new automaton capable of performing extremely delicate movements. (*Jerome McCavitt, Carnegie-Mellon University*)

An interesting capability of the T3 is its ability to monitor itself for problems. Unlike many robots that will either continue to operate as before or turn off and wait for a human to find out what went wrong if they hit a snag, the T3 can cope with certain problems that may occur during its operation. Called the "conditional" routine by its manufacturers, the T3, upon coming across a problem, will initiate two optional software (programming) sequences called "abort" and "utility." If the problem lies with the part that the robot is handling, the "abort" sequence will cause it to let go of that part along the safest possible pre-programmed route (so that the object it was working on will not be damaged). If, on the other hand, the problem is with the robot itself, the "utility" sequence will let the robot "know" that something's wrong, and it will then take the appropriate corrective action (up to, in some cases, actually fixing its own broken part).

TELEOPERATING

One type of advanced robotic worker that is very different from the two just mentioned is General Electric's Diver Equivalent Manipulator System (DEMS). This interesting machine, which is essentially a diving bell with a large robotic arm attached, is meant to aid in the search for offshore oil and gas; additionally, it will be used in salvage, construction and rescue missions. According to GE, the DEMS can easily perform all the tasks that a human diver can, including drilling, cable cutting, debris removal and inspections (through a television pickup).

On one level, the DEMS is simply an extension of the human operator. The technician controls the machine through a master control arm that is located either in the automaton's submersible chamber or on the ship floating above. By "pretending" that s/he is actually performing the task required while wearing the control arm, the operator causes the robot's huge slave arm to mimic the motions exactly. This system not only provides a much safer environment for the human involved in the work (the DEMS could take the place of several divers), but also allows much easier handling of extremely heavy equipment—the robot greatly amplifies the operator's strength.

Actually, these types of manipulators are not that unusual. What is different about the DEMS is that the operator "feels" the action of the machine through an electronic feedback system. In other words, if the DEMS is busy picking a 65-pound drill off the ocean bottom, the technician monitoring it from the ship above will feel as if s/he is in the act of picking up a five-pound weight.

Whether the DEMS is indeed a robot is open to speculation. It is actually in the class of machines called teleoperators (discussed further in Chapter 4)—sophisticated automatons that, through the use of computerization and advanced electronics, extend human capabilities in some way. On one level, it could almost be considered part of a robot—once the human "brain" joins with the DEMS "brawn" the two become a unified system, an electronic/biological robot.

Even with the advanced technology exhibited by such robots as the T3 and PUMA, there are two abilities still in the development stage that seem foremost in manufacturers' thoughts: sight and touch. Of course, as mentioned in the previous chapter, there are already a few seeing robots on the market today. "For instance," says Philippe Villers, "we are already selling a vision system that is an inspection robot in the meaningful sense of the word, but it has no arms. All it has is a TV camera and a computer system and some electrical outputs that tell you whether the part is good or bad and which part it is. And then that can be used to do anything from activating an air jet to blow away the offending part to shuttling parts on different tracks, very much as is done on railroads where cars are shuttled off on different tracks." Somewhat more elaborate vision systems, such as those that Carl Ruoff described, are being used with the Unimate PUMA and other high technology robots.

INTO THE FACTORY

While automatons with effective sight are now becoming a reality, automatons with effective tactile (touch) sensors are in the same relationship to today's robotics as we are to, say, putting astronauts on the planet Pluto—not an impossible goal, not even improbable, just a little out of our reach at present. And it would be an important step. It would be nice if a robot had the

It took this robot 12 minutes to completely weld this frame. The average time for a human worker to do the same job was about 45 minutes. (*Cincinnati Milacron*)

ability, through some kind of sophisticated sensors, to know when it had grasped a part hard enough to lift it, and yet not so hard as to break it; so that in case the vision system fails, or the object is a little less sturdy than expected, the part won't end up in pieces on the workroom floor. Robotists are now working on it, and they are developing advanced feedback systems that will, in effect, become a robot's sense of touch.

These problems are expected to be solved in a relatively short time—certainly by the turn of the century. Once they are, once the industrial robot can see, hear and feel what it's doing, once that robot is totally independent within its environment, perhaps then we will be ready for the totally automated factory that many science fiction writers have dreamed about—workplaces almost silent except for the hum and clacking of the machinery, and the steady, monotonous rhythms of the blue collar robots performing their assigned tasks.

Or perhaps not. Robots, whatever their efficiency, may always need human supervisors. Certainly they will need human programmers and, for now anyway, human repairmen who are ready with the occasional oil can. "The automated factory that people talk about is much like the Holy Grail," smiles Villers. "It's something you keep on approaching and searching for, and never quite reach."

After exploring the various advantages of a few of the high technology systems, it must be admitted that there is some difficulty in making these automatons sound exciting to those not in the field of industrial robotics. To most of us, the differences between, say, the T3 and one of Cincinnati Milacron's less sophisticated robots are not really all that meaningful. The number of types of jobs that robots can do, what kind of environment they need to do them in and how much room they take up in the workplace are all factors that may be quite important to the manufacturers who use them, but may not seem to have much to do with our daily lives.

Slowly but surely, however, robots are having a definite influence on the economic structure they are entering: an influence, some say, that will be hardly less important to our culture than the industrial revolution of the last century.

At that time, about a hundred years or so ago, American

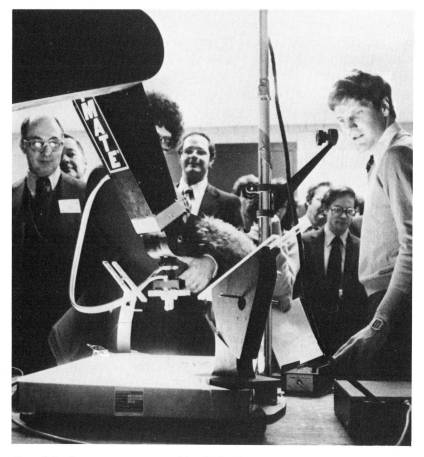

One of the first automatons capable of selecting a specific part out of a group of objects through the sense of sight. (*Jerome McCavitt, Carnegie-Mellon University*)

society was predominantly made up of farmers and small merchants. This agrarian culture was introduced to the factory and the assembly line. Multitudes of workers migrated to the cities, attracted by higher wages and the promise of an improved lifestyle; many small farmers, unable to compete with larger, industrialized farm complexes, were forced to do likewise. More and more mass-produced goods were needed to support this new population, which meant more and larger factories, which meant an influx of yet more workers. . . . The entire economic structure and, by extension, the entire cultural makeup of our society was irreversibly changed by this new technology. And now, in this century, another new and much more advanced type of technological factor is being added to the equation: the industrial robot.

The first to be affected will probably be the very people who are now working on the assembly lines. People like the auto workers who, in addition to weathering the financial turbulence and uncertainties of today's automotive industry, also have to face the possibility of being replaced with an animated metal claw.

WHO GETS REPLACED?

According to management, there is really little or no problem at all. "Robots enter the work force rather gently," asserts Mr. Engelberger. "Of our 3000 [robots], I don't know of anywhere someone could say, 'Hey, that machine took my job!' Now, of course we displace labor. We have to do that economically, but the way to displace labor so you don't have a crisis is to do it at the normal attrition rate.

"The attrition rate in metalworking in the United States is about 16 percent a year. That allows for wanderlust, pregnancy, retirements, what-have-you. So you put robots in a job that either didn't exist, or you put them in jobs that retirees leave, or you put them in jobs that are so miserable that people don't want them.

"I look to robotics as another form of automation, a way to revolutionize manufacturing. You know, 47 percent of the labor force was in farming in 1870; by 1970 only four percent of the labor force was in farming. So in a hundred-year period the entire picture has completely changed. It gave this country tremendous strength."

One of the major points that employers emphasize when talking about robots replacing labor is that the automatons are being used, for the most part, in jobs that are either very dangerous or excruciatingly boring. "What are you replacing?" asks

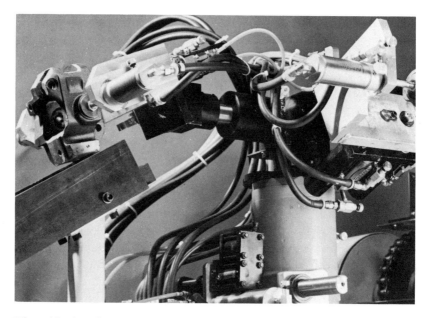

Who said robots have to resemble people? This robot has two arms—located at 90 degree angles to each other. (*Auto-Place, Inc.*)

Mr. Villers. "You are replacing people who are being used *as* robots—that's all robots are able to do."

"With robotics," agrees Engelberger, "you'll have a steady and continuing decline in blue collar workers. For 25 years they'll largely be replaced by knowledge workers. Knowledge workers are those whose human adaptability and brainpower are essential, and more economic, and a wiser thing to use than the limited amount of brainpower you get with a robot."

On the other side of the issue are the unions. For the most part, they have not opposed the addition of robots in the workplace—when they make the work safer for the workers, they are welcome; and as long as no jobs are endangered, the unions are happy to cooperate with management's desire to increase production through new technology. The introduction of this same technology means that American manufacturers will be able to stay competitive in the world market, and therefore fewer workers will be laid off.

But now that robots are becoming more sophisticated, and able to do more complicated tasks, the unions are becoming a little more wary. As is quite logical—after all, the prospect of becoming a "knowledge worker" is not extremely comforting to a 40-year-old assembly line mechanic because he probably doesn't feel all that adaptable.

At a 1979 Skilled Trades Conference on Collective Bargaining, Irving Bluestone, vice president of United Auto Workers, addressed just this issue: "The introduction of automation will be so rapid in the next five to 15 years that it will mean the kind of job displacements we must pay attention to in an ever more effective way. . . .

"A robot today costs about $20,000; that's less than what it costs to have one employee on the assembly line for one year. With such a low cost, you can anticipate the kinds of investments that will be made in GM, Ford, Chrysler and the independents to bring robots and computerized equipment into the plants."[3]

Robert Lund, a senior research associate and a lecturer in mechanical engineering at the Massachusetts Institute of Technology, recently began a long-term study of the effects of new technologies on American industry and labor. He is going to be studying various firms that have recently introduced new computerized methods (including, to a great extent, robots) into their workplaces and will then gauge the results.

Lund believes that while the technology of robotics has the capability of being extremely beneficial, we may be going about it the wrong way. "The new computers or the smart automation or the intelligent machines have characteristics that are significantly different from those in machines we are used to," he says. "Their properties tend to change rather drastically the nature of the job, the nature of its effect on people, and so you have consequences that affect everyone—managers, workers, engineers and maintenance mechanics—that have not really been thought through very carefully and that may present their own set of problems. . . . There is a push for new technology that has tremendous potential, but may not be realized unless we examine how it should be used, how it should be put in place and what the possible consequences of that technology are.

"Machines change the way people do their jobs. This is true not just for the blue collar workers, but also for engineers and management. When you change the way in which a person does a job, where the person does the job, or you change whether the

Today's robots are slowly becoming capable of taking over the production of the world's manufactured goods. Are we capable of adjusting to the change? (*Unimation, Inc.*)

person does the job or not, each situation has an impact that creates problems, and probably creates resistance to the change."

Whatever the consequences for the American worker and/or the American businessman, American (and European and Japanese) robots have become firmly entrenched in the competitive world of manufacturing. Today they are welding, painting and carrying; next year they will be assembling radios and inspecting parts. And after that? Well, perhaps in the future the term "man-made" will be practically obsolete, with the exception of a few craftsmen and artists. Perhaps, in the very near future, the guy under the hard hat will be a robot.

Chapter Four

Robots in Space

IN A SENSE, the history of space travel is the history of space robotics. With the exception of a few manned space ventures, such as the Apollo moon landing, all of our ambassadors into space have been artificial metal constructs. Through a mixture of human and computer intelligence, these relatively small craft have enabled Earth's population to experience the wonders of the solar system at second hand.

It is difficult to say exactly when our early spacecraft changed from simple automatons to robot explorers. Certainly, due to the vast distances involved in even short hops beyond the atmosphere, a certain amount of preprogrammed movement was necessary in the simplest ships. However, for the most part, this movement was limited and was in response to orders generated from Earth.

THE FIRST STEP

For example, Rangers 7, 8 and 9 were the first spacecraft to send us close-range photographs of our moon. For those photos, scientists depended on an automatic system to turn the cameras on 13 minutes before the Rangers impacted with the moon's surface. (Even these automatic controls weren't al-

The Voyagers are small, roboticized spacecraft which are exploring planets that humans are not yet capable of reaching. (*NASA*)

together foolproof—the Ranger 6 mission failed because the camera was accidentally triggered early, and it shot all its pictures before it got close to the moon.)

The first truly robotic (or at least semi-robotic) spacecraft was probably Surveyor. On June 1, 1966, it was the first Terran vehicle to make a fully controlled soft landing on the moon—all prior flights by both American and Russian spacecraft had crash-landed and were thereafter useless. Surveyor had on board a great deal of elaborate scientific equipment, all of it triggered by electronic signals from earthbound humans. While most of its movements were, again, under constant human control, the very fact that it was able to land on alien soil without anyone at the "helm" was testimony to its engineering and design.

As our spacecraft continued to venture further and further into the solar system, the sophistication of onboard equipment continued to improve. Devices for photography, measurements and taking samples (which had to be tested on the spacecraft, since there was no way to send them back to Earth) were worked on and improved. All their functions were totally automated and, like Surveyor, responded only to orders transmitted from Earth. However, as the technology performing the experiments was slowly improved upon, the equipment that drove the craft began to be capable of more independence.

When the Apollo 11 astronauts landed on the moon on July 21, 1969, many people assumed that this was the way we were going to do it from now on. This was Buck Rogers brought to life; this was humanity's destined role as explorer and conquerer of the galaxy finally come true. As a consequence, when the astronauts continued to putter around on the lunar surface for a few more missions, and then gave up the exploration of space to NASA's automated spaceships, the public lost interest. After all, how exciting could a planet landing be if there were no real men there to plant a flag and utter a few historic words?

But not everyone lost interest.

JOURNEY TO THE RED PLANET

For many years, the planet Mars had been the object of a great deal of conjecture, in both literary and scientific circles. The orb was known to have some sort of atmosphere; there was the possibility that it had moisture; mysterious lines, thought by some to be canals, were visible on its surface. Could those factors indicate the possibility of some sort of Martian life? There was really no way of finding out without actually going there.

The Apollo space mission to the moon was perhaps the last in a very long time to take humans to the surface of another astral body. There are no plans at present to send astronauts to other planets. Instead, we will send robots. (*NASA*)

Our robotic envoy to Mars: the Viking lander. Two of these spacefaring automatons touched down on the Martian surface in the summer of 1976 and sent back word on the makeup of the planet's soil. (Pictured here is a model of the actual craft.) (*NASA*)

Unfortunately, there was no chance of sending men to Mars as they had been sent to the moon. For one thing, the distances and technological difficulties were far greater; for another, the funds to build the needed equipment were just not available. So NASA settled for the next best thing—robots.

On the morning of July 20th, 1976, in an area named Chryse Planitia, the first terrestrial visitor to another planet (the moon, strictly speaking, is not a planet) set foot—or, rather, metal leg—on Mars. The second of the two Viking landers joined it, some distance away, on September 3rd.

The Viking landers were actually part of two robotic teams. Each team consisted of an orbiter that scouted out the proposed landing area from space, and in addition performed various upper atmosphere tests, and the lander.

After traveling more than 300 days in space, the Viking 1 spacecraft went into orbit around Mars. Photos that it sent back indicated that the original landing site chosen was too rough, so an alternate site was chosen, and the Viking lander ordered down.

Once scientists back on Earth gave the landing signal to the craft, the Viking was, essentially, on its own. Because of the distance between Earth and Mars, the time lapse between our sending the appropriate orders and the spacecraft receiving them would have been much too great for effective action. For example, if the Viking lander had gotten in trouble while descending, by the time the people back on Earth had figured out the problem and radioed back a solution, the lander would have been a heap of twisted scrap sitting on the Martian plains. So the robot—and at this point we can justifiably call it a robot—had to be capable, and proved itself most capable, of adjusting itself to any unknowns during its trip through the Martian atmosphere, and of computing a perfect landing on the Martian soil. This intricate maneuver involved monitoring and adjusting its angle of entry, and opening its parachute and firing its retro-rockets in the right sequence and at the right moments.

Once on the planet's surface, the Vikings were still capable of some independence. Each was equipped with a guidance control and sequencing computer that determined the lander's

A panoramic photo of Mars taken by Viking Lander 1. The tall structure to the left is the high-gain antenna that receives orders from Earth; the long arm to the right was used to gather soil samples. (NASA)

actions, either through Earth-originated commands or through programs stored in advance.

While general research was constantly in progress, the obvious question uppermost in everybody's mind was the possibility of life on Mars, however primitive. The landers contained the means for biological experiments that were assigned by scientists to be carried out by the robot (many of which consisted of "offering" the Martian soil various liquid-contained nutrients, and watching to see if anything ate them). In addition, the craft had television cameras that sent back a myriad of photos of the surrounding area and instruments for soil analysis, checking for "marsquakes" and measuring the atmosphere, wind velocity and wind direction. A long sampling arm, that was for the most part directed from Earth by remote control, scooped up soil samples and deposited them in the craft for experimentation.

Near the end of the mission, these two robots sent back a great deal of valuable information concerning the nature of the planet Mars. It was also reluctantly conceded that the possibility of life (at least in the areas where the Vikings landed) was slim.

As of this writing, the Vikings are still on Mars and still sending

A close-up view of the Viking lander's arm. There is a scoop at the end to gather samples with. (*NASA*)

back information (although their orbiters have long since ceased operating).

A RUSSIAN PIONEER

The first mobile space robot—mobile in the sense that it moved over solid ground rather than through space—was built by Russian scientists who were concentrating on getting a moving robot on the moon about the same time that we were concentrating on getting a man there.

On November 17, 1970, at 9:28 A.M. (Moscow time), the Lunakhod 1 lunar rover rolled out of the Luna 17 moon probe in which it came. This adventurous little vehicle—it measured 63 inches by 87 inches, and weighed in at a mere 1667 pounds—covered six and a half miles during its lifetime on the moon, which lasted 322 days. The rover itself was really relatively light; much of that poundage was equipment. It was run by electric motors, and solar cells were used to recharge its batteries when needed. If one of the eight wheels became stuck in the lunar soil, the scientists who designed the rover had a very simple solution—they would use a remote control signal to burst the driveshaft of that wheel, and the remaining wheels would carry the vehicle on. (The disadvantage of that scheme was that eventually they could—and did—run out of wheels.)

Actually, most of the actions of Lunakhod 1 were remotely controlled by Soviet scientists, who were monitoring the craft through its television viewers. The machine did have some autonomous functions—for example, it was programmed with a certain angle tolerance to prevent it from overturning. In other words, if the rover went over a large rock and began to tip, it would automatically give itself a stop command and wait for orders from Earth to correct the problem. Thus, it made certain decisions, and could, in that case, be called the first robot to effectively explore an extraterrestrial body.

Its cousin, Lunakhod 2, landed about three years later, on January 15, 1973. This vehicle was somewhat heavier—1848 pounds—and could travel twice as fast as its predecessor, but it wasn't much more independent.

While these enterprising robots were not very intelligent in themselves, they did demonstrate a very important lesson for the future—that roving vehicles were a practical and, in fact, desirable method for exploring the other planets in our solar system.

A STEP FURTHER OUT

Meanwhile, NASA was exploring further and further into space, and using more and more sophisticated spacecraft for that purpose. The most advanced of these explorers have been the two Voyager probes that were launched in August and September of 1977. Both these craft were targeted to fly past Jupiter and Saturn. Voyager 2 held the possibilities of also reaching Uranus and, if it was still in working condition, of sending photos back from Neptune.

Are the Voyagers actually robotic spacecraft, or just very highly automated systems? According to Carl Ruoff of JPL, "The Voyager doesn't actually pick up rocks and stuff, but it does have a lot of discrete functions—it aims its camera and does a certain amount of onboard navigation, and is able to tumble itself around to use stars to keep itself oriented. The thing is very remote. We can't get up to fix it and we can't control it directly. Not only that, it takes so long for radio signals to propagate to [arrive at] the spacecraft that we couldn't possibly control it from the ground—it takes too long. The time delay is too great, so that we wouldn't be able to make the spacecraft respond.

"So in that regard, you could say that both Voyagers were very much robotlike craft."

As the Voyagers sail on out of our solar system, let's go back to Mars for a moment since that's the last planet where we actually touched ground. Many scientists assert that we have no definite proof that there is absolutely no life on the red planet. After all, the only contact we've had there were with two immobile craft that couldn't really look very far—much as if an alien race had dropped two probes in the middle of the Sahara, and decided on that evidence that there was no life on Earth to speak of. Still, we can't send astronauts to investigate Mars as "easily" as we sent them to the moon.

For one thing, the distance is quite a bit greater: at its closest point, Mars is about 34.6 million miles from the Earth (as opposed to a distance of 239,000 miles from the Earth to the moon). This means that it would take our fastest ships about a year to get there. During that year, a manned space vessel would have to

An artistic representation of the Voyager spacecraft approaching Jupiter, its array of cameras and sensors ready for action. (*NASA*)

carry enough food, water, life support and medical equipment, and other necessities to maintain several people through more than two years (remember, they also have to get back). As of this writing, no human has stayed in space for that long a period, even just in orbit. What effect would it have physically on the astronauts' bodies, not to mention the possible psychological effects of spending that long a time in a cramped environment with literally no place to go? Then there is all the sophisticated equipment that would be needed to maintain them while on the planet's surface—and all the fuel that would be necessary to push the weight of everything just mentioned.

The above does not even take into consideration a big factor, especially where NASA is concerned: money. All of that equipment, plus the fuel needed to propel the spacecraft, would be enormously expensive. We simply cannot afford it.

OUR ROVING REPORTER

On the other hand, a robotic craft capable of landing on Mars and conducting an investigation of the planet's surface would be eminently practical. Consider a few of the advantages: a robot does not need oxygen, sleep, food, water or any of the other small necessities our astronauts would require. It can survive in extreme temperatures. It would not get bored, cramped or irritable—and it is virtually impossible for a robot to get cabin fever as astronauts might (except, perhaps, if a couple of circuits crossed). And, as with the industrial robots, an intelligent automaton would only be inconvenienced by a hurt limb—and we would only be inconvenienced by the robot's loss.

So the idea of a Martian rover is a highly desirable one. But we cannot simply use another form of the Russian Lunakhod and send it to Mars. Because of the distances involved, it would be almost impossible to send to Mars (or any planet beyond) a working robot that does not have some ability to make decisions on its own.

A round-trip communication to and from Mars takes from nine to 40 minutes, depending on the Earth's orbital distance to that planet at the time—an interval that could prove awkward if, say, an exploratory robot suddenly came across a new form of life or a large hole and wanted to know what to do about it.

The Martian rover, when completed and sent to Mars, will be invaluable in exploring the planet's surface. In this artist's conception, the rovers are gathering information and relaying their findings to an orbiter, which in turn sends the results back to Earth. (*JPL/NASA*)

Additionally, because of the time factor, a vehicle that had to wait for directions every few minutes would actually be operating only four percent of the time, while a relatively independent robot rover would be operative a good 80 percent of the time. This not only provides for more efficiency, but it saves a lot of Earth-type frustration as well. "Say the rover was moving forward and it saw a ditch," explains Ruoff. "If you waited forty minutes, the thing would already be in the ditch. So you have a

The rover as it is now being developed at the Rensselaer Polytechnic Institute in New York. This is the single-laser model: the laser bounces off the mirror at the top of the mast. When finished, the rover will carry a small computer and scientific equipment where there are now only bare circuits. (*Rensselaer Polytechnic Institute*)

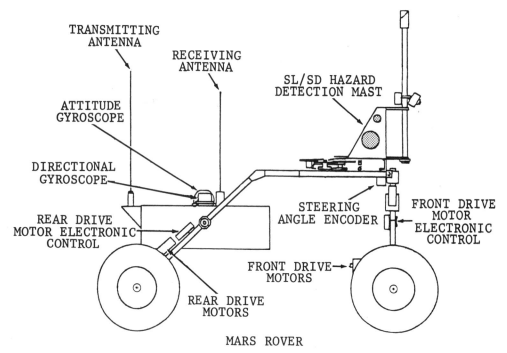

MARS ROVER

A simplified diagram of the Martian rover as it is now being developed for RPI.
(*Rensselaer Polytechnic Institute*)

choice: You can either move it an inch at a time, and wait forty minutes between each move, and it would take you forever to do anything; or you can let the thing go at some reasonable rate with a certain amount of autonomy.'' He laughs, ''But if the Martians came out with guns and said 'Your money or your life,' then we'd want it to call back—that would be reasonable.''

In addition, the highly trained, highly paid personnel who monitor the rover's activity back on Earth could turn their collective attention to more important things than telling the robot that it might be dangerous to go over the side of a cliff. ''Currently, a lot of the sequence generation is done on the ground,'' says Ruoff. ''That is, a bunch of people get together and decide what precise sequence of maneuvers the spacecraft should go through when it gets to where it's going. We're in the process of trying to automate that; so when we want to, say, take pictures, instead of having to decide precisely and at a very high level of detail where we need to aim the scan platform and when to snap the picture, we would just be able to say, 'Take a picture of so-and-so,' and the robot would do that automatically."

NASA's ideal Mars explorer would be a very autonomous rover. Since 1967, NASA, with the help of the Jet Propulsion Laboratory, the Rensselaer (N.Y.) Polytechnic Institute and several other institutions, has been working on the research and development of such a Martian explorer—a robotic vehicle that, once on Mars, would be able to wander around a large area with a minimum of human help. Right now, the rover is still in its developmental stages. The test robot being used by the Jet Propulsion Lab to try out its various systems is about the size of a Volkswagen Beetle. It is a four-wheeled vehicle with an electromechanical arm, two television cameras and various other sensing devices. While the present model uses JPL's computer for its functions, the actual rover will have its own sophisticated onboard computer.

EYEING MARS

Although a great deal of the testing equipment that will go with the Martian rover has been in existence for several years (much of it was present on the Viking landers), a lot of work still remains to make the vehicle itself as efficient and fool-proof as possible—after all, a repairman won't be able to make house calls all the way to Mars.

One of the elements still unperfected, and one that is as vital to an autonomous robot on Mars as to an assembly line robot in Detroit, is vision. Two systems are now being worked on. One is a stereo correlation system, much the same as the robotic vision described in Chapter 2 (brought, of course, to a much higher level). The second is a hazard detection unit using lasers.

Both vision types have their drawbacks. The stereo correlation system uses the contrasts between dark and light objects and their backgrounds to not only tell the computer what an object looks like, but to give its coordinates in three dimensions rather than two. This particular process is still in its earliest developmental stages. Basically, a robot can see in three dimensions by using two cameras instead of one, in much the same way that we use our two eyes to see in three dimensions—but it's not easy.

"It's very hard computationally," Carl Ruoff, who has been working on this system at JPL, explains. "Each camera feeds a scene to the computer and the scenes are somewhat different because the cameras are separate, like the difference between what your left eye sees and what your right eye sees. So you have to correlate. You've got a point in the left camera and a point in the right camera and they're not located at the same position in the scene that each camera sees, and yet they're the same physical point in space. That's called stereo correlation, and unfortunately that takes a great deal of computer time. One reason we are developing the laser system is because it is able to tell us stereo information directly."

Tom Robinson, a student at the Rensselaer Polytechnic Institute, helped develop the laser vision system for possible use with the rover. "There are two problems that we deal with," he says. "One is hazard detection, the finding of the obstacles; and then hazard avoidance, getting around them. In the early part of the game, we were really dealing more with hazard avoidance. Once we find the obstacles, how do we get around them? We really didn't know how to find where they were. We were just dealing with getting out of their way. So all the vehicles that were built had a human operator.

"In 1974 we started working on hazard detection to actually find obstacles. If it's [the rover] driving along and it's approaching a rock, we want it to be able to see the rock before the vehicle gets there and then be able to avoid it."

The RPI researchers decided that the best way to create such a hazard detection system was to use a laser, an instrument that focuses light rays into a high-intensity, coherent (extremely concentrated) beam.

The first method they tried utilized a single laser. The laser pulsar (so-called because of the "pulsing" of the signals), mounted on a flagstaff-like mast, would reflect the laser signal off of a mirror at the top of the mast, sending it toward the ground at a predetermined angle. The signal would bounce off the ground and back toward a special detector at the base of the

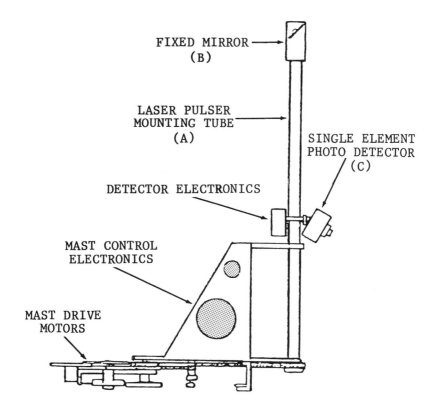

FIXED MIRROR
(B)

LASER PULSER
MOUNTING TUBE
(A)

SINGLE ELEMENT
PHOTO DETECTOR
(C)

DETECTOR ELECTRONICS

MAST CONTROL
ELECTRONICS

MAST DRIVE
MOTORS

SINGLE LASER/SINGLE DETECTOR HAZARD DETECTION MAST

The laser pulsar (A) sends bursts of laser light towards a fixed mirror (B) which reflects the light out in front of the robot. If the terrain ahead is level, the signal will bounce back at the exact angle to be picked up by the photo detector (C). However, if the ground ahead is too high or low, the angle of the returned signal will change, and it will hit either above or below the detector. When that happens, the robot "knows" that there is something in the way, and swerves to avoid it. (*Rensselaer Polytechnic Institute*) rover's mast. If the detector received the signal, it would know that the area in front was clear of obstacles and it would order the rover to proceed toward its destination. However, if the laser bounced off a large rock, or a deep ditch, the angle of the signal's return would change, and it would miss the detector. When that happened, the vehicle would know that an obstacle of some kind was in its path, and would begin to detour. While all this was happening, the mast would be rotating back and forth constantly, so that the laser signal would cover a wide area of ground in front.

The single-laser system just described was not without its problems. "With the system," Robinson says, "the vehicle couldn't distinguish between a ten-inch-high step at 1.5 meters, and a 30°slope, because both of them would make the laser shot land outside the detector's cone of vision. And *both* the step and the slope would be considered hazardous—which they're not. The vehicle can drive on a slope, but it can't get over that step." A lot of time would be unnecessarily wasted by a rover constantly detouring around small obstacles because it misinterpreted the problems they presented.

As a possible solution, the crew at RPI decided to try using several lasers with several detectors stacked one on top of the other, so that the robot would be able to build itself a sort of visual picture of the surrounding terrain.

"Let's say you have a laser shot going out at 45°," explains Robinson. "And there's just a flat floor. Chances are the middle detector will see it. Now, if you were all of a sudden to come upon a wall or a rock or something, that means that the same laser shot that went out at 45° is now going to land up a little bit higher. Maybe the fourth or fifth detector from the top will see it, rather than the detector in the middle. So, knowing the angle that the laser shot went out at, and which detector picked up that shot, you're able to get an elevation on the obstacle. Where before you just had a yes/no return, as far as determining whether there was a hazard, now all this information goes into the computer and it's analyzed, and a decision has to be made as to whether the terrain is passable or not."

As you might expect, the laser detection system is not the perfect answer either. "The laser at any given time can illuminate one pixel," explains Carl Ruoff (remember that a pixel equals a picture element). "It's aiming someplace, and it zaps the laser and reads the distance, and then it moves a little bit over and zaps again and reads the distance. So to build up a three-dimensional map of the local environment with the laser, you have to physically move the laser to point at a lot of points in the scene. And it takes a long time."

In the end, all this work will ensure that when the Martian rover finally finds its way to Mars, it will be able to go where it's been told to go with as little fuss as possible. "Basically," says Ruoff, "whether you're doing stereo correlation or you're doing laser ranging, you end up with a topographical map of the environment. The computer software doesn't care if you've gotten the map from a laser or from a vision system. It says, 'Okay, you've got a topographical map. What's in it?' " And, once the computer knows what's in the map, how many cliffs and craters and other obstacles surround it, it will be able to go on about the serious business of exploring the surface of Mars.

WHAT LIES BEYOND THE ROVER?

Next on the agenda are more exploratory vehicles, such as the nuclear-powered Galileo spacecraft, that will be launched from Earth orbit using the space shuttle, and will explore Jupiter and some of its moons. However, farther in the future (and consequently, a little more uncertain in its probability) is a much greater utilization of the resources of space and, consequently, much greater use of robots.

For any increased exploration of space, sophisticated robotic spacecraft are a necessity. "There seem to be big cost constraints," asserts Ruoff. "People would like to get more science and more information back from the spacecraft, but the missions are so complex that we can't afford to have people run them. You've seen, when we did the moonwalk down in Houston, all the people sitting in front of terminals controlling that thing. That was terribly expensive. A third of the overall cost of

The proposed Gallileo spacecraft, which is presently scheduled to be launched in January, 1982, and to arrive at Jupiter three and a half years later. It is here pictured releasing a probe which will descend into the planet's atmosphere. Afterwards, the craft will go into orbit around Jupiter for about 20 months, during which time it will send back detailed information about Jupiter and some of its moons to Earth scientists. (*JPL/NASA*)

the mission is the people who sit there and run it. So there's a lot of interest in making the thing autonomous [independent] so you don't have to have all these expenses and so the [spacecraft] can respond more when you're trying to do interesting things."

JUST ABOVE YOUR HEAD

Not all robots we dispatch into space will be exploratory. Actually, many of them will just be extensions of the type of satellites that are orbiting our world today. Satellites transmit radio and television communications, map weather conditions, study changes in global temperature and radiation levels. Most of us are unaware of the tremendous number of samples of advanced technology circling over our heads. These same satellites could, with only a few improvements, be legitimately called robots.

Right now, most of the information that the satellites pick up is transmitted to Earthside computers for analysis. However, with the advances being made today, it won't be long before those same satellites will be processing their own data, deciding which information is important, and sending only that down to Earth—live to your living room, if you have the right antenna and decoder. A March 1980 NASA/JPL study group on machine intelligence and robotics, made up of some of the top scientists in their fields, reported as follows:

"Global service robots orbit the Earth. They differ from exploratory robots primarily in the intended application of the collected data . . . . Present developments in artificial intelligence, machine intelligence and robotics suggest that, in the future, the ground-based data processing and information extraction functions will be performed onboard the robot spacecraft. Only the useful information would be sent to the ground and distributed to the users."[4]

Among these orbiting robots are some whose purposes are not as benign as simply watching the weather. Both the United States and the Russians also employ satellites as orbiting spies, carefully monitoring any military activity that may take place around the globe. As these satellites grow more and more adept at their jobs (already they can almost read a newspaper from a hundred miles up), both nations are preparing anti-satellite devices that may, someday, lead to a sort of space war, with robotic combatants exchanging laser fire high above their respective nations.

"It turns out that military spacecraft are actually quite vulnerable," states Carl Ruoff. "A lot of military spacecraft are controlled from the ground and an awful lot of our navigation and communications is done via spacecraft. However, that means they're extremely vulnerable to hunter/killer satellites and that kind of thing. From a military standpoint, there's going to be a lot of interest in how you would make a spacecraft more autonomous. So if a ground station were knocked out, the [satellite] could accommodate that, and it would still be able to do its job. Or if a sub-system on the spacecraft itself were to fail, it certainly would have to reduce its capability, but it still would be able to make a judgment about what it ought to accomplish."

ENERGY FROM THE SUN

However, the great majority of space robots will be used for more peaceful means. One of the most highly anticipated occupations of automatons will be in the construction of large structures in space. Solar power satellites, for example, will have a considerable effect on the present energy situation. Many ecologists feel that one of the most promising new energy sources today is solar power. By using the sun's rays in concert with available technology, many scientists believe that we could greatly supplement, or even replace, the world's other lagging energy resources. It is, for the most part, cheap, safe and (at least for as long as there *is* a sun) abundant.

However, the small amount of energy that is presently obtainable through the use of rooftop solar energy cells cannot come near to solving our energy problems. In order to be really effec-

Many people today believe that humanity's future lies in space, and that robots will be the key to making that dream a reality. Here, robots are assembling a solar power station which will send energy through the atmosphere to an Earth power station. (*NASA*)

tive, huge solar collectors, ranging over miles of land area, must be used—and this arrangement could backfire, seriously upsetting the delicate balance of the environment.

Solar power satellites, on the other hand, would have the unlimited regions of space in which to spread their huge collectors. The energy obtained could then be directed in concentrated beams down to relatively small ground stations. And while Earth-bound solar collectors could only be effective on sunny days, the satellites would be totally unaffected by the vagaries of the weather, receiving and converting the sun's power almost 24 hours a day, 365 days a year.

Who will build these structures? If your first answer is people, then guess again. Think about the needs of the astronauts traveling to Mars, and then multiply those needs by several hundred for all the construction workers that would be employed on such an ambitious project.

According to the aforementioned NASA study, it would cost an estimated $1.5 million per year to properly maintain a single human worker in orbit. "Frankly," adds Ruoff, "we can't afford to orbit an army of astronauts to build large structures. For one thing, astronauts are only useful in near-Earth orbit anyway. You can't send people up in geosynchronous orbit; that would be very expensive and dangerous, too, because of continuous exposure to radiation and that kind of thing. So a really clear application for automation is in space industrialization." (A geosynchronous orbit is one in which the spacecraft or satellite follows the Earth's orbit, so that it is always suspended over the same land area.)

There are two types of robots that will be used to assemble structures in space: teleoperators and computer-controlled automatons.

Teleoperators are a sort of marriage between the better qualities of human and machine. A human worker would use a robotic system to extend his or her facilities in a number of ways. It could extend sensory input, for example, by sending a television camera into a dangerous environment. It could extend manipulative abilities by augmenting the worker's strength or, conversely, enabling him or her to do precise, miniature work too small for normal human dexterity. Robots such as these are already used in many facilities and laboratories for jobs like handling highly radioactive materials. Technicians slip their hands into sets of "gloves" and, through a series of electronic relays, their movements are exactly imitated by the machine.

These, however, are only extensions of their human operators. Are they robots? "In this context," conjectures the

Robotic arms reach out to receive cargo at this imaginary solar power satellite facility. (*NASA*)

NASA study, "the term robot can then be applied to the remote system of a teleoperator, if it has at least some degree of autonomous sensing, decision-making and/or action capability."

The second type of robot used in space industrialization would be kin to many of the robots now in existence: a combination of the industrial automaton and the Martian rover. This robot would be almost totally independent; it would only need occasional human involvement, either for supervision or to correct a difficult problem. Such a robot, according to NASA, would be capable of the following: calibration (checking the correctness of a measuring instrument), checkout, data retrieval, resupply, repair, replacement of parts, cargo and crew transfer (shades of HAL in *2001*!) and recovery of spacecraft. Quite a talented entity—and very useful.

These space robots would be working on more than solar power satellites. There will also be factories that will produce goods more easily manufactured in zero gravity or in a vacuum, and scientific laboratories with similar needs.

INTERGALACTIC MINERS

Our utilization of space as an energy source may go beyond the orbiting of solar power satellites. The moon has proved to be almost barren of useful minerals. However, there are other regions available to prospective space miners.

Between the orbits of Mars and Jupiter is our solar system's asteroid belt: a 340-million-mile circle of relatively small planetoids and boulders. Scientists estimate that there may be as many as 50,000 asteroids in the belt (only counting those with diameters of at least 1/10th of a mile).

These asteroids can be considered important for several reasons. Their scientific value is, obviously, inestimable. It is conjectured that they may have come into existence about the same time as our solar system; certainly there are many secrets they could unlock about the cosmos.

However, if you are of a more materialistic state of mind, the mineral values they may hold are of equal importance. With that many potential mines to choose from, the odds are high that quite a few will hold rich veins of valuable ores. And now that Mars and Jupiter are slowly becoming visitable neighbors, it is not unlikely that corporations will be sending robotic envoys to scout out the profit possibilities of the asteroid belt.

The first of these missions will probably use fly-by spacecraft similar to the Voyagers. Close upon their heels would come craft not unlike our Martian rover—mobile robots that would explore the alien terrain, programmed to search for minerals.

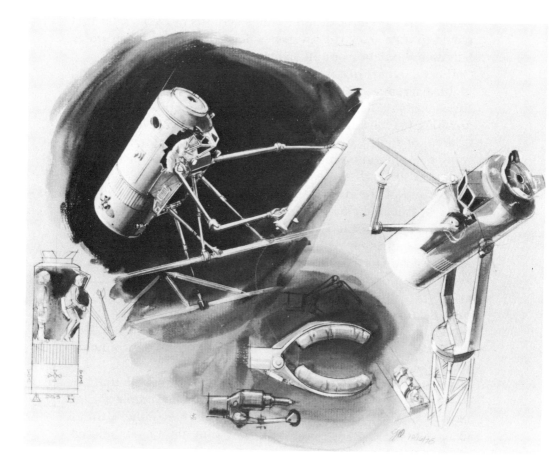

Some robots will be made up of computerized "brawn" working in combination with human "brains" to form an efficient whole. These one-man capsules would enable workers to survive in space while performing complex tasks using robotic remote manipulators. (*The Boeing Co.*)

By the time we have reached this level of robotics, it is probable that our spacecraft will have become highly autonomous. Communication between the craft and its human builders would only occur if events of particular importance demanded decisions that did not come under the automatons' programming—or if the robot was ready to scream an electronic *"Eureka!"* And, besides, these robots will be so far from Earth that, when they call home, initial contact would not be with our planet, but with large antenna relay stations in Earth orbit that would in turn relay the messages down to a central computer handling the data from all spacecraft.

That takes care of automatons that we can visualize by looking at those already in existence. But the robots that will be mining the asteroids will have to be a great deal more sophisticated than anything we have today. With perhaps only a few supervisory humans (if any) to watch, they will conceivably conduct an entire mining operation on their own, from the first tests for valuable materials through the establishing of working mines and the processing of the ores into their pure states. Finally, robotic spacecraft would carry their cargo back to a waiting Earth. All conducted quite safely, in the freezing cold of the asteroids' surfaces, entirely by robots.

Many robotists think that it could be possible for those mining workers not only to repair themselves in case of accident, but also to take the very ores they would be mining and use them to manufacture new robots. These would be called "self-replicating robots"—automatons that are capable of building their own electronic peers. An intelligent species reproducing itself—isn't that one of the criteria of life?

But let's not get into that just now. Suffice it to repeat that the future of space travel, space exploration and space industrialization is also the future of robotics. One cannot be done without the help of the other—and both may one day be considered necessities to our economical and physical survival.

Chapter Five

Show Robots

"IT'S ALMOST LIKE in the early 19th century," says Gene Beley, the head of the Android Amusement Corporation. "People having the *fun* of driving a car. My father would tell me that cars aren't fun to drive like when he was a kid. This is something where you can just see the sparkle in people's eyes."

Beley is talking about that new breed of mechanical creature popularly known as the show or entertainment robot: the humanoid-shaped, personable automatons you'll find rolling around at shopping malls and private parties. They can sing, play games, hold clever conversations—they seem, in fact, to be just as sophisticated as any robot you'll find on a movie screen. Better yet, there are no people hidden within their metallic frames—just electronic gadgetry. It's as if every robotic dream has come to gleaming life. So the question comes up: If these interesting machines can do all these things, why all the talk about the difficulty of creating robots?

The question should really be, are these really robots? Because while they highly resemble their fictional predecessors, there is one important difference: Like the scarecrow in *The Wizard of Oz*, they have no brains—at least, none they can call their own. For the most part, show robots are operated by humans through remote-control radio links; their voices are either pre-recorded tapes or are provided by that same operator standing at a discreet distance. They are large, elegant toys.

Many scientists, technicians and robotists who have spent their lives researching and developing robots maintain firmly

A typical "show robot": it looks like a robot, it acts like a robot, but is it really a robot? (*Coca Cola Co.*)

(and, occasionally, loudly) that these automatons do not in any way merit the name robot. For one thing, they do not fulfill one of the most important conditions of "robothood": intelligence. In addition, their presence may be doing serious harm to the field of robotics by raising people's hopes too high—if you think these are real robots, what respect could you have for the industrial machine just learning how to see?

Naturally enough, those who are in the business of building, renting and selling show robots see the matter somewhat differently. Anthony Reichelt, founder and president of Quasar Industries, Inc., insists that his mechanisms are, indeed, robots.

"What is a robot?" he asks. "It's a kind of euphemism, really, that was applied to a machine that has certain human qualities to it.

"Now, the industrial robots have no human qualities per se," Reichelt continues. "I mean as far as physical shape is concerned. But they do have human qualities in their work activities, and there is a definite emotional affinity or emotional recognition that's established between them and the workers that work alongside of them. If I were to give you a definition: A robot is an automated machine with the motor capabilities to duplicate some human motor functions or a shape that emulates the human shape."

QUASAR CREATIONS

New Jersey-based Quasar Industries produces several types of mobile and semi-mobile automatons. They are best known for their robot Klaatu—an example of what Reichelt calls a Sales Promotion Android—SPA for short. (The name Klaatu, as you may recall, is the name of the alien—not the robot—in *The Day the Earth Stood Still*.) These cylindrical metal men are widely seen around the country, entertaining audiences with a wide selection of jokes, wisecracks and double entendres. For the most part, they are rented out to trade shows, conventions and various other functions, but Quasar robots are also being sold to

A 5′ 9″ Andrea Android makes valiant effort to "emulate human shape."
(*Android Amusement Corp.*)

business owners who feel that a mechanical salesman that can attract customers through its mere presence is a good investment.

"We have a robot in Aurora, Illinois, for example, at Shelley's Office Supply," Reichelt says. "The robot is the floor manager. Works every day, six days a week, eight hours a day, doesn't give him any problems. There's another one in Great Bear Lake,

Quasar's Klaatu sales promotional android in typical publicity shots. Klaatu's arms and hands move, and a voice emerges from within its cone-shaped body. (*Quasar Industries, Inc.*)

Minnesota, that's a maitre d' in a restaurant. There's a robot in Charleston, West Virginia, that works for the RC Bottling Company."

Quasar Industries was originally organized because of a misunderstanding in a poll being conducted by Reichelt's marketing research firm—people thought the pollsters were asking whether they would buy a *real* (instead of a toy) robot. The poll indicated that a large number of people would pay about $3,000 for a working domestic robot, a machine that would relieve them of at least some of their household drudgery.

The idea of a domestic robot didn't sound too bad to Reichelt and his colleagues, either. However, as with many new projects, there was at least one major drawback. "We found out that it was going to cost a hell of a lot of money," says Reichelt. "More money than all of us had, or knew where to get. So we said, 'Well, look, if we're going to bet the ranch on this thing, we'd better have a robot or at least some kind of income for business that can support the research and development.' "

They decided that a promotional, remote-controlled mannikin would be just the thing, and in July, 1969, the newly established Quasar Industries finished its first robot. There were a few problems.

"I remember the robot coming through the crowd on the sidewalk," recalls Reichelt of their first appearance with the SPA. "There was an immediate crowd around him before he could even reach the door from the curb. He walked in the door, went to the left about 50 or 60 feet, and then had to go up the aisle and turn right. Well, he went up the aisle, turned left, and went crazy—destroyed a man's booth. Guy was selling nylon stockings." Reichelt laughs. "I bought $600 worth of nylon stockings that morning.

"I looked at my technicians and they looked at me, and I looked at my partners and they looked at me, and I said, 'Is this the way it's going to be?' "

Not, apparently, for Quasar Industries. At present, the company is going strong, with 32 SPA models and 20 display robots now making the rounds.

Quasar hasn't been without its share of controversy. Many robotists dislike Reichelt's insistence upon calling his automatons robots. "People will say to me, 'Some of your robots work with a remote control,' like the one in the office supply store in Aurora," he says. "The robot's in the front of the store, and there's no direct wire connection or anything, but the control and the memory and the voice of the robot are in the console in a back room. Why? The robot had to be able to be moved around the store at will, maybe change his position two or three times a day. In that particular application it was more prudent to put the computer section in a separate container. Does that make the machine any less a robot? Hell, no!

"The T3 of Cincinnati Milacron doesn't have its computer onboard. It has a cable that goes to the computer. As a matter of fact, Unimation's robot has a cable that goes to the computer and the computer runs maybe two, three or four of them. Does that make them any less robots, because a given part is not in that singular structural body? No."

And what about the domestic android that Quasar was originally formed to build? Reichelt hasn't forgotten about that, not by a long shot. "The domestic android is at a point where it's technologically feasible and possible," he asserts. "People who say that a domestic android can't be done just haven't looked at the books, or the catalogs, or the magazines, or any of the other trade journals that are readily available to the general public.

"The technology is there. It's a matter of packaging the technology, selecting the right sub-systems that are the right prices, putting them together into a reliable machine and putting it into the home, and it will do ten basic tasks."

According to Quasar's promotional brochure, those ten tasks include: answering the door, serving precooked meals, cleaning the house, monitoring for fire and unauthorized entry, monitoring heating and air conditioning, babysitting, voice recognition, running educational tapes, programming personalities for specific households and the ability to be recharged.

"The Domestic Android," reads the leaflet, "by the time it reaches the general consumer . . . will actually be a product of not only over 12 years of Research and Development, but more important will be the result of 10 years of actual Field Usage. . . .

"On a given morning, you decide that several rug-covered

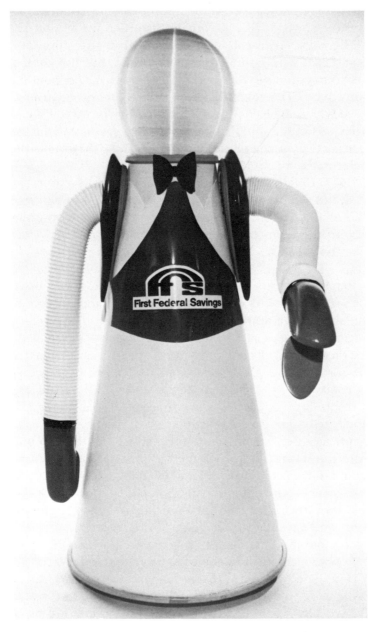

Klaatu in another of its many incarnations. (*Quasar Industries, Inc.*)

areas in your home require vacuuming. During breakfast, before you leave for work, pick-up the Quicon programmer and switch it on. Pushing a second button calls the attention of the Robot. Your Personal Servant will then be ready to receive the day's orders. By numerically inputting the date, time and function code, the Robot will confirm the selected program by replying, 'August 4th . . . vacuum rugs one and two! . . . Thank you.' You are now assured that the correct program has been selected."

Reichelt is not put off by those who say that his domestic android is impossible. "You'll get some people who will say it's impossible to build a domestic android for $4,000, or that the technology is not available. . . . Well, that's bull. All they have to do is look—it's right before you: all the technology to build robots."

Available technology notwithstanding, apparently Quasar's domestic android is taking a bit longer to perfect than first expected, as the date for it to appear on the world market was first pushed back, then indefinitely postponed. In the meantime, the company's heavy use of the SPA to promote their anticipated household android has drawn protests from both scientific and consumer circles. A good example of these claims is an article in the March-April 1978 issue of *ROM* magazine (a hobby-computer magazine) that told of a confrontation during which two of their reporters uncovered a hidden technician controlling an apparently fully independent Klaatu.

ANDROID AMUSEMENTS

While Quasar is one of the most prominent suppliers of entertainment robots, it certainly isn't the only one. There are many other companies around the country seeking to capitalize on the popularity of robots. However, their approach to the market can be somewhat different.

Gene Beley, for example, has no intentions of trying to sell a "real" robot. He is quite content with labeling his merchandise as a different breed of automaton. "I guess that's one reason why Android Amusement Corp. is not Android Corp.," he smiles. "I'm saying very, very clearly that our primary purpose has been entertainment.

"I think one of the definitions of a real robot is that it should plug itself in when the battery runs down, and in other words have some basic intelligence. A show robot, I think, is probably oriented more towards exterior design. We usually think of it as either radio-controlled or a combination of computer and radio control, and built for either television, movies or special events, where the radio control is not only the most efficient and economical way of running it, it is the most reliable way of running it under those particular conditions."

The distinction between show robots and other robotic machines does not, Beley believes, lessen the legitimacy of his product. "I like to feel that I've helped distinguish between the categories of industrial robots and what I call show robots," he says. "And I try to promote that category every time possible. I've worked with developing more and more the concept of the show robot as being a respectable entity."

Gene Beley began his career in robotics as a salesman for computerized amusement games. One day he thought it might be nice to have a robot literally bring the game to the people rather than having the people come to the game. It was then that he became interested in marketing show robots, and eventually started the Android Amusement Corp.

Incorporated in 1977 (the same year *Star Wars* came out), the AAC offers for rent and sale a variety of radio-controlled automata such as Six, a small, robotic-type mechanism, and Andrea and Adam, two humanoid "dolls" carefully sculpted to resemble typical California residents. Beley does not profess to be the owner of a fleet of independent, intelligent robots. He quite readily admits that his machines are sophisticated toys, and in fact foresees a future in which remote-controlled robots will serve just that purpose.

"Instead of being the operators behind the scenes," he says, "we've come out of the closet, so to speak, and every chance I get I hand the radio control box to the press or the public. . . . I stood on Santa Monica Boulevard in Beverly Hills and handed the

Adam Android, Andrea's 6 ' tall macho playmate, can shake its head, shake your hand, and "speak." (*Android Amusement Corp.*)

controls to a little old woman walking down the street. She never had so much fun in her life!"

Beley's latest design, the DC-1 Drink Caddy Robot, is meant to be a consumer product, to be bought somewhat in the same way that one buys a fancy new television set. The automaton, which

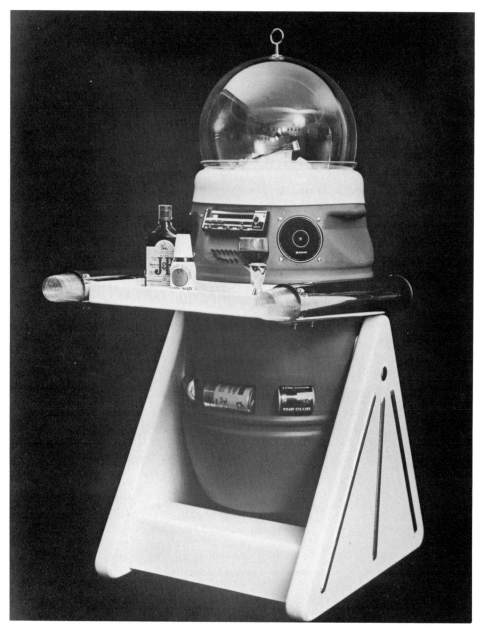

An early prototype of the Drink Caddy Robot. Wouldn't you like this rolling around your backyard? The Android Amusement Corp. hopes so. (*Android Amusement Corp.*)

is, of course, remote controlled, is actually a mini-mobile bar. Its round, translucent head is filled with ice; canned drinks are vended through the front; bottled drinks in the rear; and disposable cups are obtained through the machine's hollow arms. It also holds a tray from which guests can obtain further refreshments, and has within its body a radio/cassette unit to do its talking.

Gene Beley does not acutally manufacture his robots. They are built at The Robot Factory, a workshop located in Cascade, Colorado, and presided over by David Coleman.

"I used to ice skate with the Ice Follies," Coleman remembers. "I was doing a comedy act. The ice show was looking for a new act, so I had a meeting with the producer and all the big shots and they said, 'Well, what could you do for a new act next year?' I said, 'Well, I've already been working on it—building a robot that would ice skate.' They all laughed and thought I was nuts.

"Anyway, I went ahead and I built a robot called Ralph Roger Robot. It stood about eight feet tall. After building it, which took about two years, I auditioned it for the ice show. It ice skated, and it did spins, and it talked, and it had little lights in its arms going up and down, and the head going back and forth, and its eyes going back and forth, and it shot balls of fire out of its mouth." He grins. "That was the first robot I built."

Those ice skating robots are still doing shows, but David Coleman went on to start his own robot company. He now builds and sells automatons to people who, like Gene Beley, rent them out to conventions, shopping malls, department stores and such.

One of his most popular models, apparently, is his Six robot. "He rolls around," Coleman describes, "through remote control; he has arms that go up and down, he has pincers so that he can pick things up or can hand somebody something. He talks and he has lights in his chest and in his dome, and there are lights that flash when he talks. He has a stereo that he can play, or a cassette, or an AM/FM radio. Also, we install as an option a video camera. The camera will pick out someone or something in the audience, and then whatever it is that the robot 'sees,' or that the video camera sees, will be displayed on a monitor located in the robot's dome.

"Model Six seems to be the most popular now because it is a one-man operation and it's easily moved. In other words, it's fairly light. One person can move it up and down the stairs."

COMPUTER COPS

At this point, it may be wise to mention that not all show robots are necessarily just for show. The Advanced Robotics Corp., which puts out two "cute" robots named Argon and Huggy (which, incidentally, isn't terribly huggable), is also the manufacturer of the ESR-1: the Emergency Services Robot. This formidable piece of machinery rises 63″ high, weighs in at 585 pounds and is not the kind of thing that you'd want to meet in a dark alley. Originally designed as a fire-fighting device (it would be able to withstand much higher temperatures than a human), Advanced Robotics has now equipped it with lights, four video cameras and a highly responsive remote-control system.

Using the cameras, the mobile robot's controller can inspect a deserted factory for a distance of up to a mile from his console without endangering himself in the process. If the ESR-1 should encounter an intruder, it is capable of several types of action (all triggered by its human operator): it can temporarily blind the intruder with a high-intensity light or it can activate a siren that would effectively alarm anyone in the area and render communication with any other burglars impossible. While all this is happening (and the intruder is trying to figure out what that weird metal thing blocking his way is), the robot's operator has phoned for assistance.

Quasar Industries also markets a similar security robot, a variation on their SPA called Big Al. As Anthony Reichelt describes it, just the effect of confronting a security robot is enough to disconcert any would-be burglar. "Big Al is nine feet tall, weighs 1000 pounds and can cruise at 30 miles per hour," Reichelt says. "When he says stop, you better stop!" Put that way, there can be few doubts that security robots may prove to be highly effective in their calling.

Quasar's "Big Al" robot is meant to scare off unwanted visitors. (*Quasar Industries, Inc.*)

THE SILVER LINING

Lest it be thought that any of these men are vending their automatons strictly because of their interest in promoting robotics, it should be mentioned that they are, above all, businessmen. They have invested time and money, both in considerable amounts, in the hope that public fascination with robots can be profitable. And while they all enjoy working with their machines, idealism isn't quite enough when it comes to the actual marketing.

"I've got one friend," says Gene Beley, "who's a great robot builder and a tremendous robot enthusiast. He said, 'If I could just get $100,000, I could build this robot and really make a million dollars.' I just laughed at him and said, 'Well, a hundred thousand could get you started, maybe, but you'd be surprised how fast you would eat it up and you'd still be broke.' "

The $100,000 that Beley is talking about doesn't only have to cover the actual building of the robot, but the price of technicians, rent, advertising and marketing, and other mundane but necessary business costs.

"Prices [for a single robot] can vary anywhere from $7500 up," Coleman explains. "Now that's talking about new robots. We also sell used robots. We may sell them to someone who's not sure whether they want to be in the business or not. Perhaps they're worried about whether they can actually make a living off of a robot. So they may not be interested in putting in as much as it costs for a new one.

"It's like buying a used car. They may start out with a used one and say, 'Gee, this isn't too bad at all! You know, I could make my money back inside of a few months. So I think what I'll do is buy the new, super deluxe model, but I want to trade this one in.' So we give a good trade-in value, figuring, well, that would be a good way to get somebody else started, because naturally we're like any other company—we're looking to make a profit. As long as we can make a profit, then we can proceed to experiment and make new models and continue to make things better. That's just the name of the game, I guess."

This is not to say that the purveyors of entertainment robots

Ralph Roger Robot was created to entertain, and entertain it does: it talks, sings, and shakes hands, all by remote control. (*The Robot Factory*)

do not have dreams of their own. David Coleman, for example, is planning a show robot with voice recognition. "It can be done right now," he asserts, "but we *do* have to look at it from a business standpoint of how we can make it at least pay for itself."

"I would love to jump into having a computer in the drink-caddy robot," says Gene Beley, "and have it a lot more sophisticated. But you're talking about not only large amounts of money to develop it and build it properly—once you have it built, marketing will take an equal amount of money."

The trick, perhaps, is not to take entertainment robots too seriously. Whether or not they are indeed robots depends on your particular point of view (and, many times, on how much money you have invested in them). But the fact remains that they do fill a need in many of us—the need to be able to at least pretend that there are such things as intelligent, interesting, conversational robots.

"I think those who criticize us as saying that we're doing something naughty are taking themselves a little too seriously," Beley maintains. "I don't think they have much of a sense of humor. They're well-intentioned, intelligent human beings. They're working so hard in their particular discipline—a very pure, computerized approach to the industrial robot—that they don't appreciate our approach."

"A lot of adults will come over to me and say, 'Well, I guess the kids love that,' " says Coleman. "But if they look around, they'll see that 60 percent of the audience is not kids at all, it's adults. And they're standing right there, many of them, for a half hour to an hour, watching. It is serving some good purpose. It's entertaining."

Chapter Six

Home-Grown Robots

"I FEEL THAT the fact that there *are* robots frees us to build entire cities with our hands," enthuses Abby Gelles. "It frees you to have a less mundane life."

Ms. Gelles knows whereof she speaks. She has taught robotics to New Jersey junior high school students, and has created a curriculum for other teachers who wish to do the same. At present, there are only a few courses available in robotics around the country, just as there is no "Bachelor of Robotic Science" degree available yet at our universities. But as automatons become more and more a part of our lives, teachers and schools are discovering a gathering fascination on the part of their students in the realities of this new technology.

Actually, this interest should not come as much of a surprise. From a very young age, today's children are being repeatedly exposed to robots, both real and fictional. The science fiction films and television programs that feature robots have been tallied in the first chapter of this book; many Saturday-morning cartoons (especially those imported from Japan) also have as either their heroes or villains large, powerful robotic creatures.

In addition, children are being given an opportunity to play with what they see as robots through many toys now on the market. From the most elementary windup mechanisms that march and buzz in robot-like fashion, through larger, expensive battery-operated automatons that will walk and talk through remote control and on to the latest computerized toy vehicles

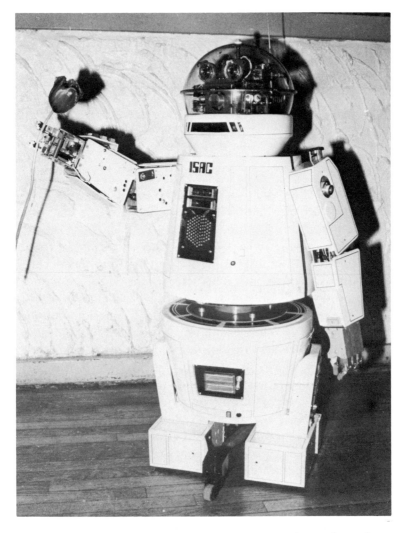

Home-built robots may not be quite as sophisticated as their industrial peers, but they can be seen as an indication of things to come. (*Courtesy Fred Haber*)

Toys such as Milton Bradley's programmable "Big Trak" are preparing children for a roboticized world. (*Milton Bradley, Inc.*)

that can be programmed by their owners to follow any desired path through the house, youngsters are learning how to live with and operate robots and their ilk.

ELECTRONIC TEACHERS

Actually, it may be more of a case of the children gravitating toward the robots rather than the robots being pushed at the

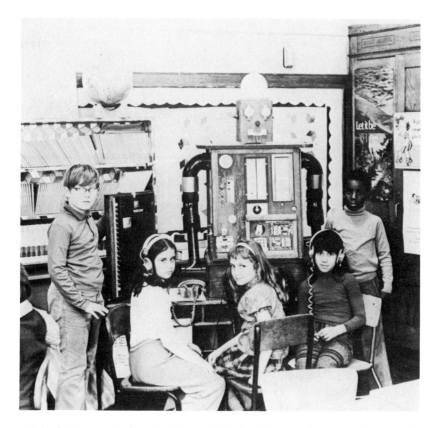

Michael Freeman's first teaching robot, Leachim, and some of its pupils. (*Michael Freeman*)

children. Educators have found that kids who are having problems with learning, or bright youngsters who are bored with the regular curriculum, are fascinated by teaching computers— which are many times presented to them disguised as robots.

One of the first of these was Leachim, a large robot/computer that was designed and built by a computer enthusiast named Michael Freeman. From 1972 to 1975, Leachim, which stood six feet high and weighed 200 pounds, taught children various academic skills by means of a pair of headphones (through which they would hear Freeman's recorded voice) and a telephone dial, with which they dialed answers to the questions put to them by Leachim.

Freeman's latest teaching toy: 2-XL. The little robotlike machine is very popular among youngsters, who are fascinated with a plaything that talks back. (*Michael Freeman*)

Nowadays, a typical teaching computer is not made to look like anything but a computer, with viewscreens and typewriter keyboards in plain view for use by the students. Since today's children are quite comfortable working with computers—more so, many times, than their parents—they do not need it disguised in robotic form.

However, robots are still fun to play with, and one of the first electronic learning toys on the market today is a direct descendent of Leachim. Built in the shape of a small, humanoid-type robot, 2-XL ("to excel") was introduced to consumers in 1978 as both a toy and an educational tool. The plastic device has four buttons through which a child can answer yes/no, true/false and multiple choice questions put to him or her verbally by the "robot."

Pre-programmed cassettes that cover a variety of subjects are being made available for use with the toy, and recently the Mego Corporation, 2-XL's manufacturer, has offered teachers a new series of educational tapes. According to their advertising brochure, 2-XL can, in addition to posing questions and responding appropriately to right or wrong answers, give further information upon request, limit the student's answering time, raise or lower the level of difficulty of a question (depending on the child's previous answer) and do simple calculations.

The success of this and other robotic toys already on the market indicates that children are quite ready and eager to embrace the new technologies of the '80s (including, of course, robotics). It is therefore even more puzzling that, once these same youngsters become old enough to wish to take a more active role in learning about and building robots, there are still so few resources at their disposal.

DO IT YOURSELF

In fact, the brighter children who are too impatient to wait until their school systems discover robotics are doing their own research and, through a great deal of trial and error, building the robots themselves. According to Tom Kemnitz, whose company, Trillium Press, is publishing Ms. Gelles' curriculum, "Most schools are not imaginative enough by any means to offer a curriculum like this. In fact, there has been no curriculum on robots in the schools, ever, as far as we know. And we've looked around for this kind of thing very deliberately. This is a whole area of education that's just been neglected."

Louis Steinberg, a 16-year-old student from Huntington Station, New York, is one of those who decided to literally take things into his own hands.

"I've been into electronics for a long time," he says. "It's a

Louis Steinberg's Self-Propelled Experimental Domestic Android—Speda for short. (*Courtesy Louis Steinberg*)

hobby of mine. My father's an electrical engineer, he's taught me quite a bit about it, and robotics . . . well, I'd seen an average amount of articles in newspapers and such on them. They looked kind of interesting. I just decided one day to build one [a robot] for the 1979 Long Island Science Congress. I ended up making it for the 1980 Long Island Science Congress." He grins. "It took longer than I expected.

"I decided that the best thing to do would be to build an inexpensive robot, because I've looked at how much it costs to buy those things commercially—$10,000 a unit—and I projected I could probably build one for under $300. I wanted to build one without cables or radio control, so you don't have to be standing there when it goes through its motions. I succeeded for about $290."

Louis named his completed project Speda, an acronym for Self-Propelled Experimental Domestic Android. As he admits, it was not an overnight job.

"Designing and building combined took about two years," he explains. "There were a lot of changes I had to make. I wasn't quite as realistic as I should have been at the beginning.

"The design is simple—I thought it up on my own. I'm using a tone-decoding system. I have a tape recorder inside the robot and I have programs made up in advance—one for vacuuming the living room and what-have-you—and on the tape I have tones, or frequencies if you will, taped on magnetic tape. By playing this, the robot electronically decides what frequencies you're playing, in which combinations, and responds to the frequencies. One frequency tells him to move his left wheel, one to move his right wheel—you play both, he turns both wheels on and goes straight; if you play one and stop the other, he'll turn right or left. . . ."

Actually, according to Louis, the idea of a home robot vacuuming the living room isn't as farfetched as most would imagine. "The basic idea is that, if we had a device like this—they are relatively inexpensive to make and purchase—we could have different tapes. The one telling him to vacuum the living room tells him to walk straight 2.7 feet and turn left 42°, and he would be just holding the vacuum the entire time.

"Right now," he continues, "[the robot] can turn right and left any angle or walk straight; he has a claw that opens and closes, and he can speak. To speak, I have him start up a second tape recorder with a prerecorded message on it.

"So for a demonstration, he'd walk up to me and tell me he's going to close his claw. I'd put something in his hand, he'd walk around, and then he can either tell me he's going to let go, or he can walk over and put it on the table. If he tells me before he lets it go, that gives me a chance to go over and take it from him before he lets it fall. All pre-programmed.''

Louis asserts that his command system—which is very similar to the touch tone decoding system used by the telephone company—not only works quite well, but allows for a great deal of flexibility in programming Speda.

"Basically, I can expand forever," he says, "and keep adding frequencies to give him more and more capabilities." However, there are a few problems that come with using sound as a programming device. "You have to watch out for a few things— harmonics, frequencies and such tend to screw up the circuits." (In other words, the complexity of most sounds—pure tones are hard to come by—can prove too much for Speda's relatively simple understanding.)

Louis is also experimenting with a voice recognition system for Speda by having the robot convert sound waves into recognizable (to it) electronic pulses. The sound waves that make up any verbal statement that the robot would be programmed to respond to would have a certain number of pulses. "If I use commands that basically don't sound too much the same, he might be able to distinguish between them. What I'm hoping is to change his speech tape so if the count is, say, 17 pulses, that would have been a question: 'What is your name?' When he receives the 17 pulses he decides 'Okay.' He fast forwards the speech tape a certain amount of time, and then plays it for a certain amount of time, and that would be the response to 'What is your name?' He would decide that's what he wants, and it would come back, 'My name is Speda.' ''

Even if all these ideas for a walking, talking, vacuum-cleaning robot don't work out as planned, Louis Steinberg won't mind too much. He has other things on his drawing board.

"There's one other [robot] I've started drawing up now that should be quite interesting if I can get it to go. You know those light-up boards in Manhattan? There's thousands of light bulbs on each. The way they change these lights bulbs now is they hang people down on scaffolds and they remove dead bulbs and put in the new ones.

"If you could build these signs with tracks up and down on

them, you could put in a robot that was built with a specific function. It could climb up and down these tracks and, by using a photocell as it passes each light bulb, it would detect if there was one burned out, stop, extend its claw to remove the bulb, place a new one in, and keep going. That means you could just let the robot do the job for you, not risk anybody's neck—probably attract a lot of publicity for the sign, also."

Not surprisingly, Louis plans to study electrical engineering when he gets to college, "possibly combined with business management. And possibly the field of robotics. Quite possibly."

Tod Loofbourrow, from Westfield, New Jersey, has become something of a success story among young robotists. His interest in automatons began when he was quite young. "When I was six years old," he relates, "my family went up to see the World's Fair in Montreal, Canada, called Expo '67. At that particular fair I saw a display on robots, and I was just fascinated by it. Since then I always really wanted to build one.

"I guess it was the summer of 1976 I actually sat down and started pulling out the electronics journals and the magazines, trying to get ideas. I started basically with a frame, and then just added a car battery and the power system and . . . . It was basically one step at a time. I didn't make an overall plan and then follow it. I just added one system after another."

Tod's creation was named Mike, short for "microtron," which in itself is a combination of "microprocessor" and "electronics." Mike is an illustration of sorts of increasing levels of technical sophistication.

"First of all," explains Tod, "you can control it; I myself can just sit there and wheel it around. It can go five different speeds forward, five different speeds in reverse, turn at various angles right and left. In that stage it really has no intelligence; it's just sort of a toy.

"In the second stage it becomes independent, and in that stage it moves around the room by itself. It's got a little ultrasonic transducer in front that sends out ultrasonic sound, and whenever that hits something and is reflected back, it knows there's an object in front of it. It's very much like a bat's radar. So it can essentially 'see' objects.

"So what it does is, it explores an area. It'll enter a room, move

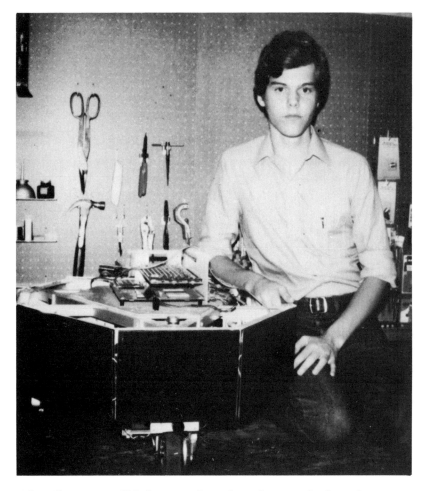

Tod Loofbourrow and his famous robot Mike. Mike proved to be such a success that Tod was asked to write a book about it. (*Courtesy Tod Loofbourrow*)

around the room, sort of finding out where the objects are with its radar; it will also bump into objects to figure out where they are (it has eight bumpers on the outside), and then it'll find the doorway of the room and go out."

Of course, with such an independent robot there is a tendency to run into problems. "Occasionally, the switches that form those bumpers get stuck, so the robot thinks it's being constantly hit by something. And it can go running across the room." Tod smiles. "I remember one time when I was demonstrating it on the stage. It thought it was being hit by one of the bumpers, and went shooting across the stage and almost dove off. Since then I've installed a panic switch on it, that allows me to take it over."

Like Louis Steinberg, Tod has built a voice recognition ability into his robot. However, his works in quite a different way.

"You have to train it," Tod explains. "You have to say the words you want it to recognize several times. For instance, if you want it to recognize 'left' and 'right,' you would say to the robot, 'Left. Left. Left.' You train it on the word 'left' until it understands how the word sounds. Then you tell it the word 'right' a couple of times, and it would store the pattern of the two words. So whenever you said something to it, it would try to see whether that matched the pattern of either the word 'left' or the word 'right.' So I can call its name and it comes, tell it to go left, it goes left; right, it goes right."

Altogether, Mike cost Tod about $450 to build; however, because the prices of electronic equipment have been dropping the last few years, he says that it can now be done for somewhere around $300.

How? Very simple. Read Tod's book. "I was asked to give a talk on it for the Amateur Computer Group of New Jersey," Tod recalls. "After the talk, a man came up to me. He thought it was a neat project, and he asked me if I wanted to write a book about it. I was sort of flabbergasted, but it sounded like fun; so that summer, I guess it was '77, I sat down and wrote a book about the robot, detailing exactly how it could be built and detailing the various systems so anyone could either build an exact duplicate of the robot or use some of the systems for their own robot design."

Tod's book, entitled *How To Build A Computer Controlled Robot*, is a 132-page, step-by-step manual detailing the building and programming of a home robot (see the Resource Index), and its publication has provided Tod with some fame in amateur robotics circles. But unlike many of his peers, and in spite of his early achievements, Tod's interests have ranged outside the areas of electronics and robotics; and he looks upon Mike as simply a hobby.

Holden Caine adjusts Herb — short for Holden's Electronic Roving Brain. Herb's mission in life is to search out the brightest light around. (*Courtesy Holden Caine*)

Other young robotists include 18-year-old Holden Caine, who has built a robot named Herb (short for Holden's Electronic Roving Brain) that will, with great dedication, look for the brightest light around and follow it, detouring around any obstacles in its path. And then there's 15-year-old Peter Quinn, who has built a simple, remote-controlled robot named Mobot. While

Peter doesn't yet have access to the type of computer necessary to be able to program his automaton for independent movement, he does not consider that an unsolvable problem. "I'll probably make the computer myself," he shrugs.

DIY* GOES TO COLLEGE

In the last ten years or so, "home grown" robots have been popping up all over America. Young people raised on science fiction and *Star Wars* are no longer content to just sit back and dream. Now that the technology is available and our knowledge of what to do with it is growing, they quite rightly see no reason why they shouldn't take advantage of it.

College students are also getting into the act on a more ambitious level. Dr. Edward Kafrissen, a specialist in microcomputers, has organized a program at the New York Institute of Technology in which students will develop a computer-controlled, industrial-type robot. Called the Adam II Project, the course is now in its first stages. Dr. Kafrissen explains:

"Adam II is a long-term project, expected to take three or four years, to build an automaton that has a degree of ability and a degree of its own intelligence. It represents several levels of sophistication above the robotic-type arms used in industry now, which are simply pre-programmed for a specific task, so that they have neither a great deal of intelligence nor a great deal of awareness of what's going on in the environment around them."

He describes the project as being part of a new concept called modularity. In this type of system, each "member" of the robot would have a certain intelligence of its own.

"Decision-making, the strategies, will be made higher up, in what you might call the brain," he says. "The brain decides what has to be done and then passes these commands on to each module, which then has enough intelligence to carry out those primitive commands. I would say the average number of com-

*Do it yourself

puters in each model will be somewhere between five and six. So it will be rather sophisticated.

"What we would hope is that people, the industrial users, could pick and choose those modules which they want. They might want modules A and B but not C and D. So by making them all compatible with the main brain, each part could be useful."

Dr. Kafrissen is quite hopeful about the eventual outcome of the Adam II robot. "I don't claim it's going to be able to do everything a human being can," he admits, "but it should be able to do tasks with some degree of intelligent adaptability of its own. We've talked to a local power company about surveillance at their power plants, especially nuclear; the possibility of using it in case of, heaven forbid, any problems. This device, of course, can go in where humans cannot because of radiation, heat, chemicals and the like." The eventual cost of such a high-level piece of robotic equipment (and, if carried out to the sophistication planned, this will be a very useful industrial tool) might be, according to Dr. Kafrissen, relatively reasonable.

And who are the professional geniuses working on this project? Well, they are neither professional robotists nor, probably, geniuses; they are undergraduate junior and senior college students. Later on, when it becomes time to work on the computerization of the robot (right now they are still involved in its more mechanical aspects), graduate students with training in computer science will be brought in.

"The science of adaptability and learning is still in its infancy," Dr. Kafrissen says, "and, in fact, some people working here nearly gave up because they felt that no machine could match a human being. But we hope that it will have enough adaptability so that if you give it a mission, it could carry it out despite the usual strange things that people handle with no trouble at all."

All these young (under 25) inventors may lead one to think that the home-built robot is the sole prerogative of the children of the space age. Not so—remember those pulp science fiction magazines of the '30s and '40s? Today's adults, who grew up on those futuristic fantasies, have succeeded in producing their own brand of household robot.

For example, Jeff Duntemann, an engineer for Xerox, constructed a metal man named Cosmo Klein that can keep track of

Arok was built by an amateur robotist who hoped eventually to put his creation on the market. (*Group W Productions*)

its own internal functions, such as hand and arm motion and voltage regulation.

Arok, a highly-publicized creation of Ben Skora, is operable either through remote control or pre-programmed tapes, and has appeared at such classy functions as Indiana's Miss Nude World contest.

Jim Lawyer, a gentleman from Tulsa, Oklahoma, made his hometown newspaper with his five-foot-high robot VAR 1. This rather artistically shaped automaton will, among other things, print out a voice analysis for interested bystanders. Lawyer hopes to be able to rent it out for trade shows and such.

Isac the robot was a team effort. From left to right: Fred Haber, Carmine Angelotti and Robert Sanborn. (*Courtesy Fred Haber*)

Interestingly, many of the adults who have created their own robots seem to have a dual purpose to their tinkering. While they are just as fascinated with the mechanics and the idea of having a robot as their younger counterparts are, there is present a somewhat more materialistic motive as well: the money that can come from robotic appearances at parties, supermarket openings and other public functions. And although they may not ever reach the prominence, of, say, a company like Quasar (or may not ever really wish to), there is no ignoring the few extra dollars a well-constructed robot can bring in.

Fred Haber, a graphic designer from Albany, New York, became interested in automatons after years of reading science fiction novels. Finally, he says, in 1978, "I got the idea of possibly building one; just trying it to see what would happen. I did some paper sketches and started coming up with ideas, bought a few extra tools for my shop, and little by little I just started putting [a robot] together. I got him to the point where he was complete, including motor, design, everything was done. But then I needed a transmitter—a remote-control radio."

So he went to a friend for help. "I asked Bob Sanborn if he would build a transmitter for me. He saw the thing, became interested in it and said, 'Gee, maybe I can get involved further.' So I took him on as an electrical engineer. And then I got Carmine Angelotti—he does script writing and public relations work and that type of thing. So now it's a three-way group."

The robot that the trio developed is called Isac, after the author who inspired its creation, Isaac Asimov. "Isac has his own voice," explains Haber. "He can be programmed to talk, to make speeches for functions, dinners, commercials. He also has a remote-controlled voice and he makes electronic sounds.

"He has arms with shoulders, elbows and wrist joints; his fingers open and close; his head turns, his wrists turn, of course he moves in any direction. . . . For now he's remotely controlled. In the very near future we're going to total programming. In other words, we'll write a program on the computer, insert it electronically and he'll go through the whole routine automatically."

Isac was first tried out at a local March of Dimes dinner, and by all accounts was a rousing success. Since then, his owners have been offering him for rent as a "show robot." However, Isac's real reason for existence can probably be found less in Fred Haber's financial status than in his strong predilection for science fiction. "I can see robots being used in the future in a lot of different areas," he says enthusiastically, "and I thought it might be fun to get in on the ground floor. The pioneering stage, so to speak."

What does all this portend for coming generations? For one thing, remember that the young people who, at 14, 15 and 16, are building simple versions of intelligent robots will, in the next ten years or so, be designing proportionately more advanced machines. Those who are not that interested in actually constructing automatons will almost certainly be either using them

Isac meets its namesake, science fiction author Isaac Asimov, the creator of the Three Laws of Robotics. (*Courtesy Fred Haber*)

or using products produced by them. At any rate, if the people in this chapter are any measuring stick to go by, robots will have a much larger role in the lives of humans than as playthings and entertainment figures.

Young filmmaker Damon Santostefano with Roublex-OMF, a model of what Damon thinks may be the household robot of the future. (*Courtesy Damon Santostefano*)

FRANKENSTEIN COMPLEX CONQUERED

But while tomorrow's adults will be prepared and probably eager for the cybernetic revolution, many of today's are not quite as ready. The sudden acceptance of robots on the part of both young and young-thinking people has perplexed both educators, who, it seems, were not really prepared for such an explosion of interest on the part of their students, and adults to whom the idea of a walking, talking, thinking machine is more frightening than fascinating. Tom Kemnitz feels very strongly that education can play a strong part in enabling both children and their parents to work with automated companions.

"In very, very practical terms, robots are the future," he explains. "We now have all kinds of people getting thrown out of work in American automobile companies who may never see work on the assembly line again, because if we ever get our assembly lines to the standards that the Japanese have, there are going to be a lot of robots doing the jobs, not a lot of people. That raises a whole series of socio-economic questions about 'What do these people do?' The answer to the challenge of robots is not to outlaw them or try not to use them. The answer is to retain human beings to do things that are valuable and important for human beings to do. Not things that machines can do."

Part of the problem, he asserts, is to abolish the fear of technology that has caused many business owners to avoid using robots in their workplaces, and has caused many workers to protest their use. Kemnitz sees the solution to these fears as a simple matter of training.

"It's very clear that people do much better with the familiar than with the unfamiliar," he says. "The familiar holds far fewer fears for them, they're much more comfortable about it, they're much less alienated from it and they're able to integrate it into their sense of who they are and where they are and what their place in the world is going to be.

"Kids should grow up being familiar with things like robots, because robots are going to be their future."

Robotics teacher Abby Gelles agrees. In fact, she says, the problem isn't really educating the children—they're already hooked. It's their parents who are in real need of education. "The kids are never afraid," she smiles. "They go home and say, 'Mommy, there's robots, they really exist! It's not just in science fiction!' That's what I'm trying to send home. And then Daddy may say, 'Yeah, you're damn right—somebody brought it into my factory and boy, what am I gonna do?' And the kid will show him the textbook, and the book has all kinds of documentation from the industry, and the book will say, 'This isn't going to hurt you. It's going to free you for life.' "

Chapter Seven

Where Do We Go From Here?

WHY ARE WE SO FASCINATED WITH ROBOTS?

QUASAR'S ANTHONY REICHELT believes that it is because of their physical similarity to the human shape. "There is a mystique about robots that people get involved with," he says, "especially those people who read, as I do, the science fiction or the science-oriented magazines. There's a thing that people get into when the machine starts looking like a human. . . . Now we've got the machine age upon us, we've got the computer age upon us, the mechanized age, and machines are starting to take the shapes of humans, that most revered shape. So you see what happens. It's subconscious."

But it must be more than a superficial similarity to the human form that makes us read about, dream about and build robots. People seem just as interested in watching a small box on wheels find its way along an obstacle course as they are in watching a life-sized mannikin wheel down the sidewalk—more, if they're the ones that built it.

Perhaps it is the fact that here is something that we ourselves have developed and created, that we can understand fully and control, that has intelligence, mobility and even a certain amount of independence. In a sense, the robot is a new and different form of mechanical life—and we are its parents.

However, like most children, the desirable qualities of robots

Motion picture robots—intelligent, independent, personable—and not very likely. (*Courtesy Lucasfilm Ltd. (LFL),* © *1980. All rights reserved.*)

and the outlook for their future depend a great deal on the people developing them.

INDUSTRIAL OUTLOOK

To industrialists, they are a new tool—a means to improve the manufacturing process in order to put out more product with less expense. Engineers and technicians from such corporations as Unimation, General Electric, General Motors and Automatix are steadily adding to the sophistication and abilities of their intelligent automatons. It is certainly no surprise that, outside of military equipment, the industrial robot is one of the most rapidly developing technologies today.

In fact, that same robot can bring together such diverse worlds as academia and industry. Carnegie-Mellon University, a highly prestigious university located in Pittsburgh, Pennsylvania, recently inaugurated what may be one of the first academic departments totally dedicated to the study and development of usable robots. The Carnegie-Mellon Robotics Institute, which officially opened its doors in December, 1980, is using its facilities to investigate and improve the state of robotics in the United States. Funded by such diverse interests as Westinghouse and the United States Navy, and employing some of the best minds in the country, the Carnegie-Mellon administrators hope to combine academic knowledge with industrial know-how to push forward the science of robotics. Besides turning out highly competent and sophisticated robots, the Institute will also soon be able to provide industry with personnel holding advanced degrees in robotics. (Most of those now working with automatons hold degrees in such subjects as computer science and electrical engineering.)

One thing to keep in mind when evaluating the usefulness of robotics is to consider some of the other areas in which the science can be applied.

For example, a great deal is being done for physically handicapped persons through the use of cybernetic limbs. For the

In spite of their utilitarian appearances, industrial robots are truly robots—
useful, independent, and getting smarter all the time. (*Jerome McCavitt,
Carnegie-Mellon University*)

blind, there are computerized machines that can "read" print
and then recite those words aloud in a typically robotic voice.

And there are other uses for industrial robots besides building
automobiles. According to Joseph Engelberger of Unimation,
"We are already working on taking one of the Puma-sized lines
and putting it under the control of quadruple paraplegics. In-

stead of trying to have that person blow air through tubes to move things, maybe we can have electrodes tied into their face to make a robot do things, so they have control below the neck.

"Or take the terrible business of having a full-time servant's attention. This way, the robot is independent, under voice communication. A person in a wheelchair can ask the robot to do things day and night, and it preserves the human dignity."

In these and other fields, the science of robotics may have far-reaching consequences, well beyond the direct use of intelligent automatons.

In April, 1980, the National Science Foundation held a workshop at the University of Rhode Island on the research needed to advance the state of knowledge in robotics. In their summary of the results, Jerome A. Feldman of the University of Rochester writes, ". . . these basic research projects have had far-reaching implications in other areas. Robotics was mainly responsible for broadening the concerns of artificial intelligence from purely formal problems into real-world activity. . . . Because of the extreme demands it places on computational constructs, basic research in robotics has led to major innovations in programming languages and systems. Robotics research is also having a significant impact on psychological and biological research on planning, manipulation and perception. It is increasingly clear that computing-based models and robotics implementations can greatly inform the study of human capacities and problems."[5]

NEXT STEP INTO SPACE

To scientists, robots are the ultimate explorers, the machines that will take us to places we could not otherwise have reached for generations. Robots have taken us into the craters of the moon, onto the sands of Mars, into the mists of Jupiter and through the exquisitely complex rings of Saturn—and they may take us further yet.

The Pioneer 10 space probe that flew past Jupiter in De-

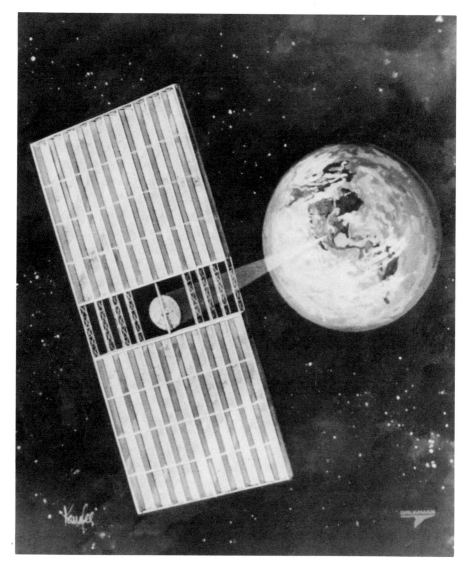

If important structures such as this solar power conversion system are to become realities, we will need robots to help build them. (NASA)

cember, 1973, has become the first Terran spacecraft to leave the confines of the solar system. It is now heading toward interstellar space at a steady and rather slow speed (on an intergalactic scale) of 25,000 miles per hour. On board the ship is a famous plaque on which is inscribed the figures of a human man and woman, various maps indicating the location of their planet and a recording of a multitude of the sounds and music of Earth.

Since Pioneer's contact with the nearest star system in its path has been estimated to be about 10 billion years in the future, it is unlikely that we'll still be around when it gets there (at least, in our present evolutionary form). But optimists are hoping that if there *is* somebody out there they may intercept the ship and interpret its message. In that case, a primitive robotic craft will have been our first contact with intelligent extraterrestrial life.

Surprisingly, there have been some complaints that our space program is not, in fact, taking full advantage of the robotic technology already in existence. In their final report, the NASA study group on machine intelligence and robotics chides the agency for not expanding its use of intelligent automation. "At rather low cost," it reads, "we could have had a remotely manned lunar explorer in progress for the past decade. . . . The Skylab Rescue Mission [referring to the demise of the Skylab orbiting laboratory in 1980] would have been a routine exercise, if a space-qualified teleoperator had been developed in the past decade. It would have been a comparatively routine mission to launch it on a military rocket. . . ."[6]

It would be unfortunate indeed if, due to lack of foresight or lack of sufficient funding, the promise of space robotics—and, in consequence, the promise of space—could not be exploited to its fullest potential.

THE CROWD PLEASERS

To the purveyors of "show androids," these machines are a new and different way to entertain people and draw crowds. There is now a demand for robots, and if it is not quite possible to supply them to the public, it is possible to give us the next best thing: a machine that *looks* like a robot.

A New Orleans photographer who recently bought his first show robot believes that the public is fascinated by these automatons because, "they're different. I mean, they like the ones

A pseudo-robot entertains an enthralled audience. What kind of real robots will exist in these children's future? (*The Robot Factory*)

with *people* inside them. They like C3P0 in *Star Wars*, they like R2D2 in *Star Wars*, and they had people inside them. And they're heroes—because this is what people think is the coming thing.

"At the last convention I did, one of the exhibitors there had a robot. To watch the crowd, how they reacted to this thing, was unbelievable. I mean, it kept people spellbound for hours!''

There is a strong possibility that, instead of remaining a publicity ploy for business concerns, the show robot will eventually become a toy for anyone who wants to build or buy one. Teenagers and adults with technical affinities are already constructing their own pet robots; in the near future, perhaps those of us with less of a scientific bent will be able to purchase full-sized, programmable automatons to play with. There is already a trend that way; for example, in the small computer-programmed automobiles that are being hawked to children, and in such automatons as the Android Amusement Corp.'s Drink Caddy Robot, a remote-controlled machine created strictly for private use.

But these will probably remain fairly expensive luxury items. As long as the humanistic robot remains out of our technical reach, people will continue to enjoy watching the colorful, chattering "androids" go through their mechanical paces. And since the builders of these machines are eager to keep up with new developments in their field, in time the show robot will become almost as sophisticated as its present-day industrial cousin. Of course, by then, who knows where the industrial and space robots will be!

TINKERERS' TOYS

To hobbyists and students, robots are shiny new toys to be fussed over, played with and constantly improved. Basement scientists and backyard tinkerers are busily developing their own answers to the question: Whither robotics? And while they have neither the resources nor, in most cases, the technical background of their professional peers, they do have incredible enthusiasm—and that can count for a great deal.

Even the head of Unimation, a very down-to-Earth manufacturer, has plans for something a little more romantic than his industrial automatons. "I fully intend," says Mr. Engelberger, "in the next couple of years, to have a robot waiter in the kitchen [of the company's conference room] that would go and make coffee and bring in food and things like that—just for show and tell. It would give people some feeling for the kind of capabilities that I would expect to come out. A maid, a butler, that everyone could have and could buy, that you could build into your house effectively and roboticize your home, take care of security, take care of fire prevention, take care of maintenance, appliances . . . all that is on."

Many scientists, however, caution that such enthusiasm should be tempered with a sense of what is actually possible. "Somebody called me up a few months ago," recalls JPL's Carl Ruoff. "He was up in Milwaukee, I think. They were going to try to build a little robot for their science and industry museum that could walk around and be able to answer questions from kids, and stuff like that. They had maybe $10,000 to build it. I was flabbergasted! I told him, 'Well, I think it's very nice that you people want to do that, but do you know how hard that problem is?'

"It's very easy to drive a robot over to pick up wrenches," he continues. "We were doing that in 1973. We could build bicycle frames and stuff like that with computer vision and robotic arms. But that didn't mean that the robot understood what it was doing.

"A lot remains to be done to make these machines smarter. There are some really interesting research issues, but in ten years I don't think people are going to be talking to robots like their little buddies. Robots will be able to do more. I don't think people will have household robots that can do a wide variety of tasks. A teleoperator is not a robot if somebody's driving it with a joystick all the time. To make a robot that can wander around in its environment and not destroy itself and accomplish useful work is a pretty good task."

HOW WILL IT HAPPEN?

The development of the robotic machine has been broken down into a series of steps. The list goes as follows: simple teleoperators, telerobots (teleoperators capable of some inde-pendence or intelligence), computer assistants, autonomous systems, full-fledged robots and self-replicating robots. We have already reached the stage at which we are producing full-fledged robots, and some scientists maintain that self-replicating robots are only some 50 years into the future.

By that time, it is hoped, automatons will be much more intelligent than they are today. They will be able to take in a variety of data about their immediate environment, study that data, evaluate it and then make decisions based on that evalua-tion without having to turn to their human builders for assis-tance. If a situation comes up that they are not familiar with, they will be able to either cope with the problem or know when and who to call for help.

Right now, it seems, we're pretty much stuck with what we've got—robots that are, for the most part, comparatively unintelli-gent, unattractive, unadaptive, unimaginative—but far from un-interesting.

After all, what challenge do the motion picture robots offer? Most of them are quite as human as their biological companions—and what is so interesting about a metallic (and usually inferior) copy of the original? Sure, they are cute—but so are puppies and small children.

On the other hand, we have the exciting reality of today's automaton—a machine in its developmental infancy, waiting to be worked on, improved and eventually brought to its full poten-tial. One day, through the combined efforts of the humans who are working with these electronic beings, we may finally pro-duce a useful, thinking, creative, artificial form of metallic life—a robot.

Resource List

GENERAL INFORMATION BOOKS

The Cybernetic Imagination in Science Fiction
by Patricia S. Warrick
The MIT Press
28 Carleton St.
Cambridge, MA 02142
(A scholarly look at robots in literature)

The Robot Book
by Robert Malone
Harcourt Brace Jovanovich, Inc.
757 Third Ave.
New York, NY 10017
(Illustrated overall look at robots)

Robots: Fact, Fiction and Prediction
by Jasia Reichardt
Penguin Books
625 Madison Ave.
New York, NY 10022
(Well-illustrated book on history of robots)

The Search For the Robots
by Alfred J. Cote, Jr.
Basic Books, Inc.
10 East 53rd St.
New York, NY 10022
(Basics of the science of robotics)

HOW-TO BOOKS

How To Build a Computer-Controlled Robot
by Tod Loofbourrow
Hayden Book Co., Inc.
50 Essex St.
Rochelle Park, NJ 07662
(Tod's book on how to build "Mike")

Tab Books (Blue Ridge Summit, PA 17214) has a series of books
on building home robots. The titles are as follows:

How To Build Your Own Self-Programming Robot
by David L. Heiserman   Tab Book #1241

Build Your Own Working Robot
by David L. Heiserman   Tab Book #841

The Complete Handbook of Robotics
by Edward L. Safford, Jr.   Tab Book #1071

How To Build Your Own Working Robot Pet
by Frank DaCosta   Tab Book #1141

PERIODICALS

Robotics Age
PO Box 801
La Canada, CA 91011
*(Bi-monthly magazine on robotics; technical and general infor-
mation articles.)*

Robotics Today
c/o Society of Manufacturing Engineers
1 SME Drive
PO Box 930
Dearborn, MI 48128

*(Magazine for those interested in robotics; emphasis on indus-
trial robots and technical information)*

Many articles on robotics can also be found in science and home
computing magazines such as Creative Computing and Popular
Mechanics.

ORGANIZATIONS

Robot Institute of America
One SME Drive
PO Box 930
Dearborn, MI 48128

*(Society for manufacturers, distributors and users of industrial
robots)*

United States Robotics Society
616 University Ave.
Palo Alto, CA 94301

*(Informational organization for the "garage experimenter";
monthly newsletter, bulletins and reprints, publishing discounts)*

EDUCATION

The Adam II Project
New York Institute of Technology
PO Box 170
Old Westbury, NY 11568

Robotics Curriculum
by Abby Gelles
Trillium Press
PO Box 921
Madison Square Station
New York, NY 10010

(Curriculum on robots for gifted junior and senior high school students)

The Robotics Institute
Carnegie-Mellon University
Schenley Park
Pittsburgh, PA 15213

Footnotes

1. *Shorter Oxford English Dictionary on Historical Principles,* Volume II (London: Oxford Press, 1973), p. 1840.

2. Karel Çapek, *R.U.R.,* translated by P. Selver (New York: Washington Square Press, 1973), pp. 17-18.

3. Irving Bluestone, "Workers' Security Crucial As Technology Accelerates," *New Technology,* Spring 1979, p. 7.

4. *Machine Intelligence and Robotics: Report of the NASA Study Group,* NASA/JPL (1980), pp. 12-13. From excerpt from "New Luster for Space Robots and Automation" by Ewald Her, *Astronautics & Aeronautics,* Volume 16, No. 9, pp. 48-60, September 1978.

5. Jerome A. Feldman, "On the Importance of Basic Research in Robotics," *Workshop on the Research Needed to Advance the State of Knowledge in Robotics,* University of Rhode Island, 1980, p. 243.

6. *Machine Intelligence and Robotics: Report of the NASA Study Group,* NASA/JPL (1980), pp. 22-23.

Index

References to captions are in *italics*.